Head-Space Analysis and Related Methods in Gas Chromatography

Head-Space Analysis and Related Methods in Gas Chromatography

B. V. IOFFE
A. G. VITENBERG

Leningrad State University
Leningrad, USSR

Translated by
Ilya A. Mamantov

A WILEY-INTERSCIENCE PUBLICATION

JOHN WILEY & SONS

New York · Chichester · Brisbane · Toronto · Singapore

© Izdatel'stvo Khimiya, 1982

All Rights Reserved.

Authorized English translation from the Russian edition published by Izdatel'stvo Khimiya.

B. V. Ioffe, A. G. Vitenberg:
Gazovaya Ekstraktsiya v Khromatograficheskom Analyze

Copyright © 1984 by John Wiley & Sons, Inc.

All rights reserved. Published simultaneously in Canada.

Reproduction or translation of any part of this work beyond that permitted by Section 107 or 108 of the 1976 United States Copyright Act without the permission of the copyright owner is unlawful. Requests for permission or further information should be addressed to the Permissions Department, John Wiley & Sons, Inc.

Library of Congress Cataloging in Publication Data:
Ioffe, Boris Veniaminovich.
   Head-space analysis and related methods in gas chromatography.

   "A Wiley-Interscience publication."
   Includes index.
   1. Gas chromatography.   I. Vitenberg, A. G. (Aleksandr Grigor'evich)   II. Title.
QD79.C45I63 1983   543'0896   83-10632
ISBN 0-471-06507-2

Printed in the United States of America

10 9 8 7 6 5 4 3 2 1

Preface

This monograph deals with a new trend in gas chromatographic analysis, which is based on the utilization of extracolumn phase equilibria in gas–condensed phase systems. An elementary theory of these analytical methods is included. Different variations of quantitative analysis, including the means of increasing analytical sensitivity and the calibration of chromatographs are presented. The application of head-space analysis to the identification and determination of individual components and substance groups present in complex mixtures, as well as to physico-chemical measurements (the measurement of distribution coefficients, activity coefficients, equilibrium constants, etc.) is discussed. Also included are the description of special instrumentation and adaptations of standard chromatographs necessary for conducting the analysis. Possible applications and advantages of the methods discussed are illustrated by such examples as the determination of volatile organic impurities in natural and industrial discharge waters, polymers, and air, and also by examples from the medical, biological, food, agricultural, and other fields.

This book is intended for analytical and organic chemists, specialists in gas chromatography, physical chemists, and biochemists. It should also be of interest to individuals involved in medicinal, sanitation, and forensic chemistry, environmental protection, and ecology. It can serve as a textbook in special courses dealing with gas chromatography, organic analysis, and analytical chemistry.

Along with the generally accepted abbreviations such as GC/MS, HPLC, and HPTLC which designate the most promising and effective variations of modern chromatography (gas chromatography-mass spectrometry and high performance liquid and thin-layer chromatography), one can observe in the current scientific and technical literature the abbreviations HSA and GCHSA, which stand for head-space analysis or gas chromatographic head-space analysis (the German equivalent is *Dampfraumanalyse*; the Russian translates as "analysis of the equilibrium vapor" or "vapor-phase analysis"). The appearance of special terminology in several languages for this unique technique of chemical analysis indicates its wide applicability and importance. Head-space analysis is of interest not only to chemists of various specialties, but also to criminologists, medical workers, and biologists. Numerous publications on the applications of head-space analysis are scattered in specialized journals in the areas of chemistry, technology, medicine, and agriculture. The purpose of this monograph is to give a reasonably complete compilation of these results and to systematically present the general principles on which they are based and the possible areas of their application. Special attention is given to new analytical techniques that utilize repeated and multistage phase equilibria and also to the description of the most recent instrumentation and technical equipment.

This book reflects the experimental work at the gas chromatography laboratory of Leningrad University, the main scientific programs of which include the development of head-space analysis, and the developments at The Perkin-Elmer Corporation, which pioneered in the development and production of equipment for head-space analysis.

The level of presentation assumes that the readers are acquainted with the principles of gas chromatography and physical chemistry (that they have a knowledge of phase equilibria). Therefore, the monograph can be used by a wide circle of readers, from teachers and students of advanced courses dealing with analytical chemistry and gas chromatography to workers at industrial testing and analytical laboratories and various government institutions.

The authors wish to express their deep thanks to Dr. L.S. Ettre, senior staff scientist at The Perkin-Elmer Corporation, for his great help and assistance in the publication of the English version of their book. It is hoped that our mutual work on this book will serve as a worthy example of productive cooperation between Soviet and American scientists.

B. V. IOFFE
A. G. VITENBERG

Leningrad, USSR
November 1983

Acknowledgments

The translator would like to express his deep appreciation to two colleagues in the Chemistry Department at Southern Methodist University, Dallas, Texas—Don E. Welsh, Instructor of Chemistry, for assistance with editing the translation, and Shirley McLean, Secretary of the Chemistry Department, for typing of the scientific text.

I.A.M.

Contents

Symbols   xiii

Introduction   1

CHAPTER ONE

Theory of Gas-Chromatographic Head-Space Analysis   9

1.1 Basic Principles of Head-Space Analysis and Related Methods   10

1.2 Measuring the Distribution Coefficients of a Liquid–Gas System   27

1.3 Static Methods of Head-Space Analysis   37

1.4 Dynamic Variations of Head-Space Analysis—Continuous Gas Extraction   49

1.5 Methods of Increasing the Sensitivity of Head-Space Analysis   57

CHAPTER TWO

Instrumentation for Head-Space Analysis   67

2.1 Basic Methods of Introducing the Vapor Phase into the Chromatograph   68

2.2	Laboratory Equipment for Gas Chromatographs	75
2.3	Automated Accessories for Standard Gas Chromatographs and Specialized Analyzers	85

CHAPTER THREE

Application of Head-Space Analysis for the Quantitative Determination of Impurities   **99**

3.1	Analysis of Water and Aqueous Solutions	100
3.2	Determination of Volatile Organic Substances in Biological Systems	117
3.3	Determination of Volatile Substances in Polymers	131
3.4	Analysis of Food Products	146
3.5	Determination of Gases in Solutions	148

CHAPTER FOUR

Equilibrium Concentration of Impurities of Gases (Reverse Head-Space Analysis)   **167**

4.1	Characteristics and Basic Types of Reverse Head-Space Analysis	168
4.2	Equilibrium Concentration in Nonvolatile Liquids	170
4.3	Equilibrium Concentration in Volatile Liquids	180
4.4	Equilibrium Concentration in Volatile Liquids with Varying Values of the Distribution Coefficients	204
4.5	Head-Space Analysis Combined with Equilibrium Concentration	209

CHAPTER FIVE

Qualitative Analysis and Other HSA Applications   **215**

5.1	Individual and Group Identifications	216
5.2	Study of the Chromatographic Profile of Complex Mixtures—Analysis of Odors	224
5.3	Application of Head-Space Analysis to the Calibration and Testing of Gas-Chromatographic Instruments	234
5.4	Determination of Ionization Constants of Organic Substances in Solutions	242
5.5	Determination of Activity Coefficients	254
5.6	Determination of the Molecular Weight	256
5.7	Head-Space Analysis in Microbiology and Medical Diagnostics	257

Index   **265**

Symbols

A	Peak area of the chromatographic peak (indices G, L, i, and n indicate the phase, component, and extraction number, respectively)
C	Concentration (indices indicate a phase or a component).
C^0	Initial concentration
C^{st}	Concentration of the standard reagent
$\bar{c}$	Constant characterizing the mass-transfer process
D	Diffusion coefficient
d	Diameter of the packing particle
d_L	Density of the liquid
F	Vapor volume of the volatile liquid in unit volume of the equilibrium gas phase
F_c	Flow rate of the carrier gas
F_k	Factor that takes into account the difference in distribution coefficients in the biological object of interest and in water
f_i	Functional relationship between concentrations of the ith component in equilibrium phases
f	Sensitivity coefficient of the chromatographic detector (calibration factor)
f_r	Coefficient that takes into account a relative detector sensitivity of the substance of interest and the standard
f_s	A characteristic of solvent volatility (in gas extraction and equilibrium concentration)
H	Henry's constant
K	Distribution coefficient
K_x	Distribution coefficient when concentration is expressed in mole fraction

SYMBOLS

K_e	Equilibrium constant
l	Column length
M	Molecular weight
m	Quantity (mass) of matter
n	Number of replacements of the analysis gas phase with a pure gas
p	Total pressure
p_i	Partial pressure of ith component
p_i^0	Vapor pressure of a pure component
P_L	Solvent vapor pressure
q	Fraction of solute in the mobile phase
R	Universal gas constant
r	ratio of volumes V_G/V_L
S	Analytical sensitivity
T	Absolute temperature
t^{min}	Time needed for removal of the substance from concentrator
t_{99}	Diffusion time of 99% volatile impurity from polymer into gas phase
u_0	Carrier gas linear velocity
V	Container volume for establishing phase equilibrium
V_G	Gas-phase volume
V_g	Specific retention volume
V_L	Liquid phase volume
V_L^0	Initial volume of volatile liquid
V_M^0	Adjusted retention volume of nonabsorbed substances
V_R^0	Adjusted retention volume
v_G	Volume of mobile phase in a column
v_G^{max}	Maximum volume of a mobile phase through a stationary phase
v_G^{min}	Minimum volume of gas permeating the stationary phase at equilibrium concentration
v_g	Volume of gas batched into chromatograph
v_l	Volume of liquid fed into chromatograph
w_L	Quantity of liquid phase in the column-concentrator
X	Concentration of solution after gas extraction in fractions compared with the initial fraction
x	Mole fraction
Y	Fraction of volume of the volatile solution compared with the initial volume after the gas extraction
Z	Rate of extraction of volatile matter during gas extraction

GREEK

α	Degree of increase of the analytical sensitivity; ratio of volumes V_G/V_L
γ	Activity coefficient
δ	Permissible error of analysis
μ	Chemical potential
φ	Volume or weight fraction of phase in the polymer dispersion
Ψ	Angle coefficient of the logarithmic dependence of the gas-phase concentration on the number of sequential extractions

Head-Space Analysis
and Related Methods
in Gas Chromatography

INTRODUCTION

During the last decade, methods of gas chromatographic analysis of volatile materials based on the application of phase equilibria outside the chromatographic column have been widely accepted. This rapidly developing trend of chromatographic analysis, which is acquiring an independent value, foresees the use of the substance to be analyzed as one of the phases of the heterogeneous system. The analytical and physico-chemical characteristics of this phase are determined by analyzing the other phase into which some part of the components of the first phase is partitioned with the establishment of equilibrium. Usually, the analyzed substance is in a condensed phase (liquid or solid), while direct analysis being carried out on the equilibrium gaseous (vapor) phase. The name of the method, "analysis of equilibrium vapor" or "vapor-phase analysis," (head-space analysis, HSA) is derived from this equilibrium condition.* There are similar methods that are in effect reverse head-space analysis. In these, the sample is a gas in equilibrium with a liquid or solid phase and the condensed phase is analyzed. There are also variations of analysis based upon the equilibria between condensed phases.

A distinctive feature of the HSA technique is that the chemical information contained in the gas phase is used to determine the nature and composition of the condensed phase with which it is in contact. It is necessary to stress that the same principle is employed by one of our sense organs—the sense of smell. Animals, by means of smell, obtain from their surroundings the information needed for food, self-preservation, orientation, and association with others. In the evolutionary path, this method of receiving information was little developed, and is much weaker in human beings than in the majority of animals. The development of HSA methods opens up wide possibilities for determining trace contaminants in the atmosphere, in other gaseous media, and in other substances. The characteristics of the HSA method make it in many cases irreplaceable and very effective.

The first characteristic is the possibility of the determination of the volatile components of the samples to be studied. The direct admission of the samples into the gas chromatograph is impossible or unsuitable due to the insufficient sensitivity of the detectors, the possibility of decomposition,

---

*However, one must keep in mind that the thermodynamic equilibrium between the phases is not required in all practical applications, and there are possible methods of determination that assume only a certain degree of equilibrium (for instance, during identification and certain qualitative determinations).

the undesirable contamination of the column by nonvolatile residue, or the danger of perturbing the chemical equilibrium of the system. Examples of HSA applicability are the widely known methods of blood analysis for alcohol content and poisonous volatile materials, the effectiveness and official recognition of which contributed to the technique of head-space analysis. Here also belong the standard methods of determination of residual monomers and solvents in polymeric materials. The sanitation-hygienic control of polymeric materials by the HSA method became an object of intense interest and received special recognition with the discovery of the carcinogenic properties of vinyl chloride, necessitating strict control of vinyl chloride content in many commodities.

The usual direct analysis of such sample types does not limit the value and possibilities of head-space analysis. Recently interesting new perspectives have appeared, which open up HSA methods for investigating chemical equilibria in solutions. If volatile agents (or products) are present, then the analysis of the equilibrium vapor permits the determination of equilibrium constants in compound mixtures without separation of their components. For instance, it is possible to determine the dissociation constants of volatile bases by means of measuring the ratios of their vapor concentrations over the solutions with known and varying pH values. It is very important that such determinations can be carried out in complex mixtures of undetermined composition when there is no known applicable method of measuring dissociation constants.

The other characteristic of head-space analysis is its relative ease of automation: leading manufacturers of gas-chromatographic instrumentation have produced special automated analyzers and devices for sampling equilibrium vapor.

Existing HSA methods (including automated variations) are based on sampling from an enclosed space in a static system. An interesting and important prospect for further development is the transition to chromatographic analysis of the gas phase in open systems, that is, the analysis of the gas flow passing through an analyzed solution.

The continuous process of gas extraction in which the volatile components of a solution are extracted by gas bubbles passing through the solution can be described by various equations depending on the experimental conditions. This ability opens up new practical and diverse HSA applications. Of the new prospects, the first application that should be mentioned is the possibility of component determinations in complex systems with unknown distribution coefficients. To such belong, for instance, stratal or sewage waters, biological fluids, and biological tissues.

In a gas extraction process, both concentration and distribution coefficients of the analyzed components can be simultaneously determined. Therefore, interesting physico-chemical applications become possible such as measurements of the distribution and activity coefficients in solutions, including the fairly important field of maximum dilutions.

Besides the analytical and physico-chemical applications, the utilization of HSA principles for testing purposes (checking and calibration of the gas-chromatographic instrumentation) should be mentioned. Applications of both static and dynamic methods are possible in this area. The static variations allow us to reproduce simply and exactly the standard gaseous mixtures based on heterogeneous systems with known as well as with unknown distribution coefficients. Dynamic methods (with application of continuous gas extraction) can produce gas flows in which the concentration of components change regularly with time. Without sophisticated instrumentation detector calibration can be obtained and their linearity over a wide range can be checked.

The first analytical works based on the HSA principle were accomplished 20 years ago. The original investigations completed during these years are summarized in three special surveys[1-3] (at present they are to a certain degree outdated) and a monograph.[4]* Methods related to head-space analysis are also discussed in the book by V. G. Berezkin and his co-authors.[5] In it the main attention is given to the problem of the identification of organic substances by their distribution coefficient between two appropriately selected liquid phases. Berezkin has termed the combination of all possible methods of analysis by combining the partitioning of the investigated substances between two phases with their chromatographic determination the "chromato-distributive method." He considers head-space analysis a special case. However, the importance acquired by head-space analysis and the number of works devoted to it considerably exceed the role of similar methods employing two condensed phases. One may assume that HSA methods soon will acquire the same importance for organic analysis that liquid extraction has enjoyed in contemporary inorganic analysis.

Head-space analysis was used in practice even in the early period of development of gas chromatography. In 1957 a Belgian thermal power

---

*See also a new review[26].

station used a semiautomatic device for continuous monitoring of the hydrogen content in boiler water by gas-chromatographic analysis of the vapor phase.** Poor solubility of hydrogen in water creates a very favorable condition for head-space analysis, and it is possible to determine the hydrogen content in the amount of 1 ppb, that is, 10^{-6} g in 1 liter of water.

The first example of the use of head-space analysis for the quantitative determination of organic substances is probably Weurman's[6] investigation on the enzymatic generation of the volatile components of raspberries, in which a linear relationship was noted between the heights of the chromatographic peaks in the vapors of aqueous solutions of the simplest oxygen compounds and their concentration (10^{-2}–10^{-3}%). By changing the technique of selecting the samples, Bassette, Özeris, and Whitnah[7] were, by this method, able to determine the volatile compounds in even more dilute solutions (up to 10^{-4}–10^{-6}%). They noticed the considerable differences in the sensitivity of such analyses for various classes of compounds and also the increase in the slope of the straight lines when changing from the simplest to the more complex members of a homologous series[8].

Even before the development of gas chromatography, in 1939, the possibility and expediency of using analysis of the vapor phase for the determination of alcohol content in aqueous solutions and urine were indicated.[9]. The successful application of head-space analysis to the determination of alcohol content in blood, first described by Curry and co-workers,[10] contributed to the widening acceptance of this method and promoted the development of appropriate instrumentation.[11]

MacAuliffe[12] advanced the concept of the analysis of systems with unknown partition coefficients by multiple gas extraction, and various applications of head-space analysis with discrete gas extraction were studied in detail at Leningrad University.[13,14]

Wahlroos[15,16] was the first to consider the use of continuous gas extraction for the analysis and concentration of volatile impurities and the determination of partition coefficients. However, his works remained unnoticed, and twelve years later various possibilities of using gas extraction not only from nonvolatile but also from volatile solvents were again indicated.[17]

The reverse variation, equilibrium concentration of microimpurities of gases on chromatographic stationary phases, was proposed independently

---

**L. Bovijn, J. Pirrote, and A. Berger, in *Gas Chromatography 1968* (Amsterdam Symposium) D. H. Desty, ed., Butterworths, London, 1958, pp. 310-320.

by Czechoslovakian[18] and American[19] analysts, and equilibrium concentration in liquids without limitation of their volatility was studied in the gas chromatography laboratory of Leningrad University.[20]

Indicative of the increasing interest in head-space analysis is the recent organization of special colloquia and symposia dedicated to the problems of HSA techniques; two such international colloquia were conducted by the Perkin-Elmer Corporation in Überlingen (FRG)[21,22] and in the fall of 1977 a symposium was held in Chicago under the auspices of the American Chemical Society on the analysis of food and beverages, the works of which were published in book form.[23] The papers read at the 2nd International Colloquium in Überlingen and at the seminar on head-space analysis held in October 1978 in Beaconsfield were also collected in a published volume.[24] The first Soviet seminar on head-space analysis was held at the beginning of December 1979 at Leningrad University with the participation of about 150 chemists from 24 cities.[25]

REFERENCES

1. B. Kolb, *Angew. Gas-Chromatogr.*, (Bodenseewerk Perkin-Elmer and Co., GmbH, Überlingen) **15**, 1 (1972).
2. A. G. Vitenberg, B. V. Ioffe, and V. N. Borisov, *Zh. Analit. Khim.*, **29**, 1975 (1974).
3. B. Kolb, *J. Chromatogr.*, **122**, 553 (1976).
4. H. Hachenberg and A. P. Schmidt, *Gas Chromatographic Head-space Analysis*, Heyden, London, 1977, 125 pp.
5. V. G. Berezkin, V. D. Loshchilova, A. G. Pankov, and V. D. Yagodovskii, *Chromatoraspredelitel'nyi Metod*, Nauka, Moscow, 1976, 112 pp.
6. C. Weurman, *Food Technol.*, **15**, 531 (1961).
7. R. Bassette, S. Özeris, and C. H. Whitnah, *Anal. Chem.*, **34**, 1540 (1962).
8. S. Özeris and R. Bassette, *Anal. Chem.*, **35**, 1091 (1963).
9. R. N. Harger, E. G. Bridwell, and B. B. Raney, *Proc. Amer. Soc. Biol. Chem.*, *J. Biol. Chem.*, **128**, xxxviii (1939); quoted in R. N. Harger, B. B. Raney, E. G. Bridwell, and M. F. Kitchel, *J. Biol. Chem.*, **183**, 197 (1950).
10. A. S. Curry, G. Hurst, N. R. Kent, and H. Powell, *Nature*, **195**, 603 (1962).
11. D. Jentzsch, H. Krüger, and G. Lebrecht, *Angew. Gas-Chromatogr.*, (Bodenseewerk Perkin-Elmer, Überlingen), **9**, 8 (1967).

REFERENCES

12. C. McAuliffe, *Chem. Technol.*, **1**, 46 (1971).
13. A. G. Vitenberg, B. V. Stolyarov, and S. A. Smirnova, *Vestn. LGU*, **16**, 132 (1977).
14. B. V. Ioffe, B. V. Stolyarov, and S. A. Smirnova, *Zh. Analit. Khim.*, **33**, 1296 (1978).
15. Ö. Wahlroos, *Ann. Acad. Sci. Fennicae, Ser. A, II. Chemica*, **122**, 1 (1963).
16. Ö. Wahlroos, *Acta Chem. Scand.*, **20**, 197 (1966).
17. A. G. Vitenberg and B. V. Ioffe, *Dokl. Akad. Nauk SSSR*, **238**, 352 (1978); B. V. Ioffe and A. G. Vitenberg, *Chromatographia*, **11**, 282 (1978).
18. J. Novák, V. Vasák, and J. Janák, *Anal. Chem.*, **37**, 661 (1965).
19. A. Dravnieks and B. K. Krotoszynski, *J. Gas Chromatogr.*, **4**, 367 (1966).
20. B. V. Ioffe, A. G. Vitenberg, and V. N. Borisov, *Zh. Analit. Khim.*, **27**, 1811 (1972); A. G. Vitenberg and B. V. Ioffe, *Dokl. Akad. Nauk SSSR*, **219**, 921 (1974).
21. *Colloquium über die Gas-chromatographische Dampfraumanalyse*, Kurzfassung der Vorträge. Bodenseewerk Perkin-Elmer, GmbH, Überlingen, 1975, pp. 20–21.
22. Vorträge zum 2. Internationalen Colloquium über die Gas-chromatographische Dampfraumanalyse in Überlingen (October 18–20, 1978), Bodenseewerk Perkin-Elmer, GmbH, Überlingen, 1978.
23. G. Charalambous (Ed.), *Analysis of Foods and Beverages. Head-space Techniques*, Academic Press, New York, 1978, 394 pp.
24. B. Kolb (Ed.), *Applied Head-space Gas Chromatography*, Heyden, London, 1980, 186 pp.
25. B. V. Stolyarov, *Zh. Analit. Khim.*, **35**, 1653 (1980).
26. J. Drozd and J. Novàk, *J. Chromatogr.*, **165**, 141 (1979).

CHAPTER ONE

Theory of Gas-Chromatographic Head-Space Analysis

1.1 BASIC PRINCIPLES OF HEAD-SPACE ANALYSIS AND RELATED METHODS

The principles of all methods of quantitative head-space analysis are based on the thermodynamic conditions of the phases. These conditions require the existence of certain interrelationships between the concentrations of each of the components in different phases of any heterogeneous systems that are in equilibrium. In particular, the equilibrium distribution of the substance between the coexisting phases of each of the i components in each of the j equilibrium phases is described by the equality of chemical potentials:

$$\mu_i^1 = \mu_i^2 = \ldots = \mu_i^j \qquad (1.1)$$

But the chemical potential of the ith component in the jth phase is a function of its concentration C_i^j in the given phase (more correctly, of its composition). Thus, it follows from (1.1) that a certain dependence must exist between the concentrations of each component in equilibrium phases I and II.

$$C_i^I = f_i^{I,II}(C_i^{II}) \qquad (1.2)$$

Relation (1.2) can serve as the basis for establishing the composition of one of the phases (i.e., for finding C_i^I) by means of analysis of another phase (i.e., determination of C_i^{II}). This is a distinguishing feature of the methods discussed in this book. The indices of the function $f_i^{I,II}$ reflect the dependence of the numerical relationship of C_i^I and C_i^{II} on the nature of a given component i and the correlation with phases I and II. It is necessary to take into consideration that the values of $f_i^{I,II}$, which are a function of the composition (nature) of the phases, depend also on such parameters of equilibrium as temperature and pressure (for the condensed phases, this dependence is less pronounced), and on the method of expressing the concentration.

Generally, for analytical purposes, it is not necessary to establish the shape of the f function, because one may use empirical standardization by plotting computational graphs or tables under strictly specified conditions. However, for the majority of modern HSA applications for the determination of impurities in low concentrations, the simplest case of functional dependence $f(C)$ acquires a special importance—a direct proportionality between concentrations of the given component in the equilibrium phases:

$$C^I = KC^{II} \tag{1.3}$$

The coefficient of proportionality K is called the *distribution coefficient* and is determined by the same factors that influence the values of the function $f(C)$, which are given above. The simplest law of distribution (1.3) is valid in very different heterogeneous systems, independent of the aggregate condition of phases, yet considerably limited with respect to the universal relation (1.2) by the effect of the concentration dependence of the chemical potentials.[1] Expressing the chemical potentials of the components through their mole fractions x_i and activity coefficient γ_i,

$$\mu_i = \mu_i^0 + RT \ln (\gamma_i x_i) \tag{1.4}$$

and then solving for phases I and II in terms of (1.1), one obtains

$$\mu_i^{0I} + RT \ln (\gamma_i^I x_i^I) = \mu_i^{0II} + RT \ln (\gamma_i^{II} x_i^{II}) \tag{1.5}$$

or

$$\ln \frac{\gamma_i^I x_i^I}{\gamma_i^{II} x_i^{II}} = \frac{\mu_i^{0,II} - \mu_i^{0,I}}{RT} \tag{1.6}$$

The right-hand side of this expression, which contains the difference of standard values of chemical potentials of the ith component in the first and second phases, is a constant (at given temperature, pressure, and method of selecting a standard condition). Therefore, the relationship of activities $\gamma_i x_i$ in different phases also must be constant.

$$\frac{\gamma_i^I x_i^I}{\gamma_i^{II} x_i^{II}} \equiv K_a \tag{1.7}$$

The ratio of the mole fractions x_i^I/x_i^{II} will retain its integrity only if the activity coefficients remain uniform (γ_i = const.), as in maximally diluted and ideal solutions. In the latter case, $\gamma_i = 1$ over the entire interval of concentrations 0 to 1. Thus, the distribution law takes the form

$$x_i^I = K_x x_i^{II} \tag{1.8}$$

and will be used without concentration limitations only in heterogeneous systems in which the phases can be considered as ideal solutions. In real solutions the consistency of the ratio of concentrations x_i^I/x_i^{II} is maintained as $x_i \to 0$. Although expressed by various methods, the concentrations of diluted solutions are proportional to each other. Hence, Equation (1.3) with the concentrations given in any units is equivalent to Equation (1.8). When calculating the composition of such solutions, one can use not only mole fractions x_i, but also any convenient method of expressing concentrations (with the corresponding values of K). Criteria concerning the concentration limits within which Equations (1.3) and (1.8) remain linear are fairly uncertain, as they depend on the nature of the system and the sensitivity of the analyses (e.g., x_i limits may be indicated on the order of 10^{-2}). In any case, formula (1.3) is a reliable basis for the determination of trace amounts of impurities (<0.01%) according to the recommended range of the analyzed concentrations[2].

It is conventional in the HSA method to consider the liquid (or solid) phase whose composition must be determined as phase I and the gas phase as phase II. All values of the coefficient K that appear in this book will follow this convention. The concentrations of the component being determined in the gas and liquid phases will be designated by the indices G and L, respectively.

To the gas–liquid systems belongs the first empirical regularity in phase equilibrium—Henry's law* (a tenet established at the beginning of the 19th century, long before the advent of chemical thermodynamics). In its original form, Henry's law stated the direct proportionality between the solubility of gases in liquids and the pressure p (at constant temperature):

$$C_L = K_p p \qquad (1.9)$$

The proportionality constant in this formula is similar to the distribution coefficient K of Equation (1.3). However, it differs in that it is not dimensionless but has the dimension of inverse pressure. The present form of Henry's law expresses the direct proportionality between the partial pressure of a volatile component i and its mole fraction in the liquid phase:

$$p_i = H_i x_{i,L} \qquad (1.10)$$

H_i, Henry's constant, is inversely proportional to the distribution coefficient K_x of Equation (1.8). Certainly, at pressures affording ideal gas behavior, the mole fraction of component i in the vapor phase is numerically

---

*William Henry (1775–1836), the author of *Elements of Experimental Chemistry*, is also known for his works on the oxidation of ammonia and the composition of gases in the dry distillation of wood, peat, and coal.

1.1 BASIC PRINCIPLES OF HEAD-SPACE ANALYSIS

equal to the relationship between the partial pressure p_i and the total pressure p (the pressure observed at the coexistence of phases) and therefore

$$p_i = p x_{i,G} = H_i x_{i,L} \tag{1.11}$$

Compared with (1.8) this gives

$$H_i = \frac{p}{K_x} \quad \text{or} \quad K_x = \frac{p}{H_i} \tag{1.12}$$

Henry's constants are used frequently in practical thermodynamic computations. There exist correlations with the critical temperatures and gram-molecular volumes that allow H_i values to be determined fairly exactly.[3] However, for analytical applications, distribution coefficients are more convenient and can be computed from Henry's constants by the use of relation (1.12). The published data on liquid-vapor equilibrium usually belong in the range of average concentrations. Therefore, in establishing the values of the distribution coefficients in dilute solutions, one must use extrapolation. For these computations the logarithm of the activity coefficient of a dissolved substance, $\log \gamma_i$, is obtained by extrapolations (graphically or by means of semiempirical or empirical equations[4]), and then Henry's constant is computed:

$$H_i = \gamma_i^\infty p_i^0 \tag{1.13}$$

where γ^∞ is the limiting activity coefficient (extrapolated for the maximum dilution) and p_i^0 is the vapor pressure of pure component i at the temperature of the phase equilibrium.* The distribution coefficient K_x can be computed from the ratio (1.12)

$$K_x = \frac{p}{\gamma_i^\infty p_i^\infty} \tag{1.14}$$

In many cases the exactness of such computations is not very high. The results obtained by the various methods of extrapolation in very polar liquids diverge from each other and from the experimental data for dilute solutions by as much as several tens of percent.[5] For these systems the theoretical computations of γ_i^∞ and K_x based on molecular-statistical models give the same low precision, insufficient for analytical purposes.[5]

---

*Formula (1.13) follows from Henry's law (1.10) and the relationship $p_i = \gamma_i^\infty x_{i,L} p_i^0$, which, when $\gamma_i = 1$, becomes Raoult's law.

For practical purposes, the best method is to use the experimental values of the distribution coefficients in dilute solutions, measured by vapor-phase analysis as given in Section 1.2. It should be mentioned that an important physico-chemical application of the analysis of the equilibrium vapor is the determination of the activity coefficients of the volatile substances in solutions of maximum dilution according to the coefficients K and H_i:

$$\gamma^\infty = \frac{d_L RT}{K p_i^0 M} \qquad (1.15)$$

d_L represents the density of the liquid solvent, M is its molecular mass, and R is the gas constant, which is equal to 62,370 when pressure is expressed in millimeters of mercury and volume in milliliters or to 8.3143 kJ/(kmol · K).

Formula (1.15) is derived indirectly from the distribution law, Equation (1.3), with the assumption that the vapor phase conforms to ideal gas behavior. The properties of maximally diluted solutions practically conform to those of pure solvent. It might be noted that concentrations are expressed in weight-volume or molar-volume units. The molar-volume concentration of the solute is equal to the product of its mole fraction x_i and the number of moles of the liquid phase expressed in volume units. For practical purposes, the density and average molecular weight of solutions of infinite dilution are equal to d_L and M, respectively. Hence, the molar concentration of the substance in the liquid phase can be written as

$$C_L = \frac{x_{i,L} d_L}{M}$$

The molar-volume concentration of the solute in the vapor phase is equal to $x_{i,G}/V_{G,M}$, where $V_{G,M}$ is the molar volume of the vapor phase. $V_{G,M} = RT/p$; thus

$$C_G = x_{i,G} \frac{p}{RT} = \frac{p_i}{RT}$$

but considering the generalized Raoult's law

$$C_G = \frac{\gamma_i^\infty p_i^\infty x_{i,L}}{RT}$$

the ratio of the concentrations C_L/C_G gives

$$\frac{x_{i,L} d_L}{M} : \frac{\gamma_i^\infty p_i^0 x_{i,L}}{RT} = \frac{d_L RT}{\gamma_i^\infty p_i^0 M} = K$$

which is equivalent to (1.15).

Table 1.1 is a summary of the data concerning the distribution coefficients in dilute solutions at various temperatures. These results were obtained in the gas chromatographic laboratory of Leningrad University. In Table 1.2, values are given for the distribution coefficients of a liquid–gas system for six compounds representing the basic classes or organic substances in 81 solvents at 25°C. These were obtained by head-space analysis.[6] The data in these tables can be used to determine the dependence of the distribution coefficients upon temperature and the nature of the liquid phase in liquid–gas systems. These dependencies are very important in both determining the feasibility of analytical determinations and increasing their sensitivity.

The temperature dependence of the distribution coefficients, which can be derived from relationship (1.6), has the form

$$\ln K_i = \frac{A_i}{RT} - B \qquad (1.16)$$

where $A_i = \mu_i^{0,\text{II}} - \mu_i^{0,\text{I}}$ is the difference between the standard values of chemical potentials of the given component in the vapor and liquid phases. The factor B includes the logarithms of the relationship of the activity coefficients of the ith component in equilibrium phases as well as the mode of conversion from mole fractions to concentration units as expressed by a mass of the component in a volume unit. A_i and B change little with temperature. If narrow temperature ranges are used, the linear dependence of the logarithm of the distribution coefficient upon the inverse of the absolute temperature is retained. In particular, the linear dependence $\ln K$ vs. $1/T$ is usually maintained in 20–30° temperature intervals near room temperature (Table 1.1 and Fig. 1.1) However, deviations from linearity were noticed for the solutions of the simplest ketones, butanol and ethyl acetate in paraffin oil.[13]

The relative temperature coefficient K_i according to (1.16) is equal to

$$\frac{1}{K_i} \frac{\Delta K_i}{\Delta T} = -\frac{A_i}{RT^2} \qquad (1.17)$$

As long as the values of A_i/R for the majority of investigated organic substances are in the range of 2300–7000, the change in K_i for 1°C (at ambient temperature) is 3–8%. In practical applications, this discrepancy must be considered. As a rule, quantitative head-space analysis must be conducted under thermostated conditions with an accuracy within a tenth of a degree.

Table 1.1 Distribution Coefficients of Organic Substances in Liquid–Air (or Liquid–Nitrogen) Systems at Temperatures from 10 to 30°C[a]

Substance	Solvent	Interval of Mass Concentrations C_L%	Temperature, °C					Method[b]	Source of Data
			10	15	20	25	30		
Hydrocarbons									
Benzene	Water	10^{-4}–10^{-5}	7.6	6.2	4.8	4.0	3.4	S	7
	65% CH$_3$COOH	10^{-4}–0.1	276[c]	146	(124)[d]	106	72[e]	D	8
	80% CH$_3$COOH	10^{-4}–0.5	684[c]	385	313	270	219	D	8
	CH$_3$COOH	10^{-4}–0.5	—	—	895	780	620	D	8
	PEG–300	0.02–1	—	—	641	601	456	D	9
	PEG–400	10^{-4}–0.7	—	—	(741)[f]	586	375	D	10
	Squalane	10^{-2}–1.2	—	—	(598)[g]	(500)[g]	420	D	10
Toluene	Water	10^{-4}–10^{-5}	8.1	6.0	4.6	3.6	2.9	S	7
	65% CH$_3$COOH	10^{-4}–0.1	602[c]	294	(236)[d]	194	125	D	8
	80% CH$_3$COOH	10^{-4}–0.5	1,715[c]	854	665	518	415	D	8
	CH$_3$COOH	10^{-4}–10$10^{-5}$	—	—	2480	2135	1760	D	8
	PEG–300	0.01–1.0	—	—	(1457)[h]	1154	1046	D	10
m-Xylene	Water	10^{-4}–10^{-5}	9.7	7.4	5.9	4.0	3.9	S	7
	65% CH$_3$COOH	10^{-4}–0.1	1,012[c]	499	(406)[d]	320	229	D	8
	80% CH$_3$COOH	10^{-4}–0.5	3,680	1,880	1580	1300	1110	D	8
	CH$_3$COOH	10^{-4}–10^{-5}	—	—	6370	5570	4570	D	8
Ethylbenzene	65% CH$_3$COOH	10^{-4}–0.1	849	398	(322)[d]	257	172	D	8
	80% CH$_3$COOH	10^{-4}–0.5	2,276	1,284	(1062)[d]	983	581	D	8

			Oxygen Compounds					
Ethanol	Water	0.01–1.0	(10,200)[j]	7,970	7020	5260	(4440)[j]	D	11
1–Propanol	Water	0.01–1.0	(9,280)	6,950	5480	4090	(3210)	D	11
1–Butanol	Water	0.01–1.0	(8,610)	6,410	4660	3600	(2710)	D	11
Acetone	Water	0.0–1.0	1,418	1,171	752	551	484	D	9
Methyl ethyl ketone	Water	0.01–1.0	1,150	810	600	380	(283)[j]	D	11
Dioxane	Water	0.01–1.0	14,840	10,340	8000	5750	(4330)	D	9
Ethyl acetate	Water	0.04–1.0	431	299	210	150	(108)	D	9
n–Butyl acetate	Water	0.01–0.7	296	189	126	87	(59)	D	9
			Sulfurous Compounds						
Methyl mercaptan	Benzene	—		178	150	124	91	S	12
Ethyl mercaptan	Benzene	—		594	486	406	336	323	S	12
Dimethyl sulfide	Benzene	—		728	563	488	438	419	S	12

[a] At atmospheric pressure; concentrations in mass per unit volume.
[b] S, static method; D, dynamic (continuous gas extraction); see Section 1.2.
[c] At 0°C.
[d] Interpolated by means of relationship (1.16).
[e] At 35°C.
[f] Extrapolated from data for higher temperatures.
[g] Extrapolated from data for 30 and 50.5°C ($K^{50.5} = 218$).
[h] Extrapolated from data for 25, 30, and 35°C.
[i] Extrapolated from data for 15, 20, and 25°C.
[j] Extrapolated from data for lower temperatures.

Table 1.2 Distribution Coefficients in Liquid–Gas Systems at 25°C

Solvent	n-Octane	Toluene	Ethanol	Methyl Ethyl Ketone	Dioxane	Nitromethane
Carbon disulfide	14,300	8,900	143	661	2,654	264
Cyclohexane	13,500	3,900	80	355	1,137	126
Triethylamine	13,300	4,450	690	677	1,692	458
Diethyl ether	13,200	6,760	2,114	1,003	2,616	1,170
Ethyl bromide	11,100	11,800	—	5,777	8,907	2,115
n-Hexane	9,900	3,520	82	355	974	135
Isooctane	9,800	2,680	61	310	886	118
Tetrahydrofuran	8,870	10,800	2,168	2,136	6,141	5,379
Diisopropyl ether	8,800	4,020	719	786	1,725	719
Toluene	8,705	—	273	1,339	3,956	1,452
p-Xylene	7,620	5,120	268	1,260	3,632	1,191
Benzene	8,310	7,930	320	1,880	6,597	1,994
Chloroform	7,150	10,500	779	9,632	35,646	2,686
Carbon tetrachloride	7,030	5,620	149	997	4,342	384
Dibutyl ether	6,920	3,750	368	592	1,586	490
Methylene chloride	6,302	11,100	632	—	3,368	4,994
n-Decane	6,302	2,571	62	265	820	114
Chlorobenzene	5,972	6,261	265	1,901	5,237	1,468
Bromobenzene	4,808	5,762	248	1,600	4,812	1,327
Fluorobenzene	4,567	7,245	336	2,339	6,361	2,460
2,6-Lutidine	3,805	4,875	3,681	1,516	3,870	2,739
Squalane	3,512	1,775	42	184	640	84

Hexafluorobenzene	3,512	8,181	—	—	4,342	1,408
Ethorybenzene	3,043	4,609	337	1,371	3,708	1,812
2-Picoline	2,945	5,174	4,705	2,035	4,685	3,778
Ethylene chloride	2,852	7,459	559	4,442	—	4,369
Ethyl acetate	2,852	5,020	1,279	—	4,946	5,179
Iodobenzene	2,852	4,567	216	1,224	4,140	1,031
Methyl ethyl ketone	2,535	5,121	1,878	—	4,812	6,994
Bis(2-ethoxyethyl)ether	2,466	4,260	1,384	1,174	3,235	4,369
Methoxybenzene	2,466	4,224	431	1,783	5,087	2,739
Octanol-1	2,339	1,895	2,168	639	1,518	411
Cyclohexanone	2,225	5,020	1,690	1,747	4,239	5,379
tert-Butanol	2,172	1,561	6,051	—	2,869	705
Tetramethyl guanidine	2,027	4,738	19,715	1,600	4,046	7,362
Isopentanol	1,982	1,631	3,527	907	1,953	603
Pyridine	1,982	5,452	5,294	2,404	5,936	5,594
1, 4-Dioxane	1,900	5,070	1,878	1,729	—	5,594
1-Butanol	1,823	2,016	4,758	1,039	—	759
2-Propanol	1,688	1,566	—	1,318	2,195	858
1-Propanol	1,657	1,902	4,705	1,085	2,403	903
Diphenyl ether	1,628	3,620	230	980	3,490	1,180
Acetone	1,599	4,121	—	2,838	5,237	8,800
Benzonitrile	1,519	4,260	926	2,307	5,936	4,994
Tetramethylurea	1,401	5,282	7,636	1,840	4,140	9,994
Dibenzyl ether	1,301	3,089	356	1,009	3,225	1,861
Acetophenone	1,282	3,899	1,054	1,646	4,946	4,112

Table 1.2 Continued

Solvent	n-Octane	Toluene	Ethanol	Methyl Ethyl Ketone	Dioxane	Nitromethane
Hexamethylphosporic acid triamide	1,264	4,447	38,539	1,747	3,632	15,724
Ethanol	1,264	2,221	—	1,490	2,695	1,499
Quinoline	1,247	3,424	2,918	1,416	3,708	2,911
Nitrobenzene	1,152	3,782	424	1,729	5,396	3,883
m-Cresol	891	2,270	12,284	24,091	99,028	2,973
N,N-Dimethyl acetamide	776	4,155	9,211	2,035	4,685	13,202
Acetic acid	750	2,075	7,061	3,331	17	3,778
Nitroethane	714	3,782	834	2,935	7,422	8,377
Methanol	608	1,487	—	2,247	3,870	2,851
Benzyl alcohol	581	1,881	3,527	2,247	8,907	2,084
Dimethyl formamide	577	3,840	6,778	2,163	4,946	13,994
Tricresyl phosphate	566	1,487	463	603	1,241	1,452
Methoxyethanol	526	1,961	2,821	1,371	3,956	5,179
Nonylphenoloxethylate	481	1,067	651	380	1,049	1,394
N-Methyl-2-pyrrolidone	459	3,960	7,368	1,662	4,342	12,607
Acetonitrile	418	2,666	1,761	2,935	8,097	13,327
Aniline	356	2,240	1,837	3,685	13,707	5,594

Methyl formamide	285	1,528	5,647	1,516	3,632	5,594
Cyanoethyl morpholine	235	1,532	936	853	2,280	3,494
Butyrolactone	208	2,799	2,286	1,901	5,087	9,994
Nitromethane	205	2,041	1,125	2,886	9,898	—
Dodecafluoroheptanol	187	837	6,051	26,282	162,000	3,778
Formyl morpholine	170	1,874	2,644	1,121	3,235	7,362
Propylene carbonate	168	1,769	1,205	1,543	4,451	8,745
Dimethyl sulfoxide	121	1,745	7,061	1,270	3,870	13,994
Tetrafluoropropanol-1	119	880	8,560	28,437	71,298	8,229
Tetrahydrothiophene-1,1-dioxide	61.0	1,248	1,407	997	3,870	7,362
Triscyanoethoxypropane	53.0	775	662	649	1,953	4,236
Oxydipropionitrile	49.8	1,010	794	1,039	3,490	6,358
Diethylene glycol	38.9	240	1,407	221	699	927
Triethylene glycol	24.6	263	1,140	201	699	869
Ethylene glycol		105	2,644	270	1,081	743
Formamide		138	3,527	857	3,295	2,739
Water			5,647	469	5,396	1,105

Source: Rohrschneider[6]

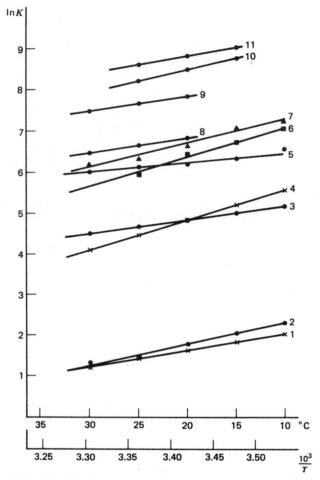

Figure 1.1. Relationship of the distribution coefficients to the temperature in liquid–air systems. *1*, Benzene in water; *2*, *m*-xylene in water; *3*, methylmercaptan in benzene; *4*, *n*-butyl acetate in water; *5*, dimethyl sulfide in benzene; *6*, methyl ethyl ketone in water; *7*, acetone in water; *8*, benzene in acetic acid; *9*, toluene in acetic acid; *10*, isobutanol in water; *11*, ethanol in water.

The distribution coefficients, greater than unity, decrease with increasing temperature, converging to a limiting value of unity as the temperature approaches the critical point. (At the boiling point of the solvent, they differ considerably from unity.) The decrease of distribution coefficients with increasing temperature is used to increase the sensitivity of head-space analysis. This effect is considerably greater in water solutions, where a

1.1 BASIC PRINCIPLES OF HEAD-SPACE ANALYSIS

temperature increase of 60°C can lead to a 10–20-fold increase in sensitivity.

The dependence of the distribution coefficients upon the nature of the liquid phase and the dissolved substance was studied until recently mainly in relation to a problem of the effective separation of substances in gas chromatographic columns.[1] Published information refers mainly to nonvolatile liquids used as stationary liquid phases. However, the possibility of applying the distribution coefficients to the characterization of solvent properties[6] led to the compilation of the considerable amount of data on organic liquids given in Table 1.2.

From Tables 1.1 and 1.2, it is evident that the numerical values of liquid–air distribution coefficients of volatile organic compounds vary over a wide range from several units to 10^4–10^5 units. High values of the distribution coefficients on the order of 5000–10,000 are observed between infinitely miscible substances of similar nature. Examples are solutions of hydrocarbons in hydrocarbons or halogen derivatives and solutions of the simplest alcohols in water. Poorly soluble components such as hydrocarbons in water are characterized by very small distribution coefficients. As the mutual solubility increases, the distribution coefficients also increase, reaching several thousand with liquids of limited solubility and in systems that have a tendency toward phase separation. The largest distribution coefficients (on the order of 10^4–10^5, and sometimes 100,000 and greater) are apparently the result of specific interaction with a solvent. Usually this results from the development of stable hydrogen bonds and the establishment of acid-base equilibrium. (Dioxane, in particular, functions as a base in systems with cresol and fluorinated alcohols.)

The dependence of distribution coefficients (or Henry's coefficients) upon the composition of mixed solvents is a fairly important question. For ideal mixtures of solvents, I. R. Krichesvskii indicated[14] that the logarithm of Henry's coefficient in infinitely diluted solutions of the volatile component must follow the additivity rule

$$\ln H_{1,2} = x_1 \ln H_1 + x_2 \ln H_2 \qquad (1.18)$$

where H_1, H_2, and $H_{1,2}$, are Henry's coefficients in solvents 1 and 2 and their mixture respectively, and x_1 and x_2 are mole fractions of solvents 1 and 2 in the mixture ($x_1 + x_2 = 1$). The distribution coefficient and Henry's constant for a given system are inversely proportional (1.12). A completely similar rule of additivity can be written for the logarithms of the distribution coefficients in mixed solvents:

$$\ln K_{1,2} = x_1 \ln K_1 + x_2 \ln K_2 \qquad (1.19)$$

or

$$\ln K_{1,2} = x_1 \ln \frac{K_1}{K_2} + \ln K_2$$

Positive deviations from the additivity correspond to positive deviations from Raoult's law in the mixture of solvents. Conversely, negative deviations from Raoult's law lead to decreases in the distribution coefficients.

In nonideal systems (e.g., mixtures of water with an alcohol, glycol, or dioxane), the deviations from the linear relationships (1.18) and (1.19) can reach 10–100% ($\Delta \ln H_{1,2} \pm 0.25$ for argon[15]), but for gases in organic solvents they do not usually exceed a few percent. A number of investigations deal with computations of deviations from relation (1.18). It is possible to compute Henry's constants in mixed solvents from the density and molecular parameters of the components. The activity coefficients of the solvents in the mixture need not be known. The accuracy of such computations for nonpolar and polar solvents (on the average) is approximately 5%.[15]

Equation (1.19) generally characterizes the effect of various mixed substances on the distribution coefficient and allows one to draw considerable conclusions concerning analytical applications. The effect of adding foreign substances is proportional to their concentration; impurities in smaller quantities can be disregarded, and this gives a reliable basis for head-space analysis of dilute solutions. However, increasing the concentration of substances the distribution coefficient of which differ sharply from that of the basic solvent can cause a considerable change in $K_{1,2}$. Such changes must be considered during the analyses. An example of this consideration involves the analysis of trace quantities of aromatic hydrocarbons in air by the method of equilibrium concentration in acetic acid. In such cases there is a considerable decrease in the distribution coefficients due to absorption of atmospheric moisture, which is present in much greater quantities than the aromatic hydrocarbons that are being determined (see Chapter 4). The magnitude of such effects in water solutions is given in Table 1.3, from which it is evident that the addition of 1% of organic solvents increases the distribution coefficient of benzene or toluene by 5–10% with acetone and by 20–50% with alcohol or ether. At the same time, the addition of salt decreases the distribution coefficients of organic substances. This salting-out effect in concentrated salt solutions is quite considerable. As concentrated solutions of soluble salts approach saturation, the distribution coefficients of alcohols, carbonyl compounds, ethers, and hydrocarbons decrease ten-

Table 1.3 Influence of the Addition of a Substance upon the Value of the Distribution Coefficient of an Organic Substance between an Aqueous Solution and Air

Compound	Added Substance and Its Concentration	Temp., °C	K In Water	K In Given Solution	Source of Data
Benzene	Sulfuric acid, 0.1N (pH 1)	15	6.6	6.7	9
	Sodium chloride, 1%	15	6.6	6.1	9
	10%	15	6.6	3.3	9
	Acetone, 1%	15	6.6	6.9	9
	Ethanol, 1%	15	6.6	9.8	9
	Ether, 1%	15	6.6	7.5	9
	Potassium acetate, 47% (pH 14)	20	4.8	0.70	8
Toluene	Sodium chloride, 1%	15	6.2	5.8	9
	10%	15	6.2	2.9	9
	Acetone, 1%	15	6.2	6.8	9
	Ethanol, 1%	15	6.2	7.6	9
	Ether, 1%	15	6.2	7.6	9
	Potassium acetate, 47% (pH 14)	20	4.6	0.57	8
m-Xylene	Potassium acetate, 47% (pH 14)	20	5.9	0.49	8
Acetone	Potassium acetate, 36% (pH 14)	25	(580)[a]	44	8
Methyl ethyl ketone	Potassium acetate, 36% (pH 14)	25	(380)[a]	30	8
Ethanol	Potassium acetate, 36% (pH 14)	25	(5260)[a]	760	8
	Sodium sulfate (sat)	28	(4805)[b]	333	13
1-Propanol	Sodium sulfate (sat)	28	(3550)[b]	145	13
1-Butanol	Sodium sulfate (sat)	28	(3020)[b]	75	13
Acetone	Sodium sulfate (sat)	28	769	71	13
2-Pentanone	Sodium sulfate (sat)	28	272	12.2	13
Isovaleric aldehyde	Sodium sulfate (sat)	28	50	4.2	13
Ethyl acetate	Sodium sulfate (sat)	28	141	8.5	13
n-Butyl acetate	Sodium sulfate (sat)	28	(68)[b]	5.7	13

[a]Data of Table 1.1.
[b]Extrapolated according to data of Table 1.1.

fold from their initial value. The pronounced decrease in distribution coefficients brought about by salting out is used, as will be shown later, to increase the sensitivity in head-space analysis.

Two groups of methods of head-space analysis can be distinguished, depending upon the conditions of the phase equilibria:

1. *Static methods*. When the equilibrium between the gas and the condensed phases forms a closed system.
2. *Dynamic methods*. When contact between the phases occurs in an open system in which the gas is blown through a layer of liquid or a granulated solid phase.

In any of the methods detailed in the following sections of this chapter, it is assumed that the functional relationship between the concentrations of coexisting phases is utilized. However, the aim of the analysis is to determine the concentration of the initial sample before its contact with the gas phase. Under such conditions the formula acquires a special value satisfying this requirement, which is derived in the following manner.

While an equilibrium is being established between a liquid sample of volume V_L and a gas that occupies volume V_G, a certain portion of the volatile substance, contained in the liquid, will move into the gas phase. The equilibrium concentration of the liquid C_L will then be less than the initial concentration C_L^0. The relationship between them will be determined by a mass balance

$$C_L^0 V_L = C_L V_L + C_G V_G$$

where, according to distribution law (1.3), $C_L = KC_G$, and thus

$$C_L^0 V_L = KC_G V_L + C_G V_G$$

or

$$C_L^0 = C_G \left(K + \frac{V_G}{V_L} \right) \qquad (1.20)$$

This formula is the foundation of all static methods of head-space analysis and it also represents the basis of the deduction of the equations describing the dynamic method, the final formulae of which are more complex, since they include exponential or logarithmic functions. The same formula also forms the basis of the more useful methods of measuring distribution coefficients. This will be discussed in the following section.

1.2 MEASURING THE DISTRIBUTION COEFFICIENTS OF A LIQUID–GAS SYSTEM

One of the number of experimental methods gaining notice in the measurement of distribution coefficients is utilizing gas-chromatographic techniques. Measuring K in this case involves the dependence of the adjusted retention volume of the chromatographic substance (V'_R) upon the volume of the liquid phase (V_L) in the chromatographic column.

$$V'_R = KV_L \qquad (1.21)$$

The value of K is calculated from the results obtained in measuring the retention parameters of the column by utilizing a stationary phase of known composition. It must be noted that the chromatographic process is useful for the determination of K in liquids with low vapor pressure (functioning as stationary phases in the gas chromatographic columns). The limitation of this method lies in the insufficient accuracy of the determination of the low values of K (<10). The general drawback of the chromatographic process in the determination of K is its possible inability to reach equilibrium and also its dependence on the retention volume of the eluting substance. Further problems are related to distribution between the liquid and gas phases and to the adsorption of the substance on the surfaces of the liquid phase and the support.

In addition to the chromatographic method, Rohrschneider[6] recommended an indirect method for the determination of K. This method is based on head-space analysis with the use of a standard substance whose distribution parameters in a studied system are known. It requires reliable data on the absolute values of the distribution coefficients or the activities of the standard substance in the solvent used.

The value of K can be determined experimentally from the basic distribution law [Equation (1.3)] after obtaining the equilibrium concentration of the substance in the liquid and gas phases. The shortcoming of this approach for determining K exists in the necessity of independently establishing the amount of the substance in the liquid and gas, which requires various methods of analysis.

The limitations of this method lie in the removal of only one phase for analysis at various concentrations of the distributed substance. A possibility is to determine the change in the concentration of a substance in one of the phases after thermodynamic equilibrium with another phase has been achieved. The basic advantage of determining K by this method is that the relative responses of a number of detectors are very reliable. Systematic errors are excluded by relating the results to the conditions of the gas-chromatographic analysis and to the sample size. Obviously, the most accurate results are achieved when determining a range of concentrations if the deviations from linearity of the distribution isotherms C_G vs. C_L are within the limits of error of the measurement.

Of the various gas-chromatographic techniques for the determination of distribution coefficients, those based on the measurement of the concentrations in one of the phases seem to be the most promising. Therefore, major emphasis will be given to them in this section.

At present, there are three variations of this method (two static and one dynamic). The first is being used to determine low values of K (10.00–50.00). This version is based on the substitution of the gas phase in equilibrium with the solution of the substance for a pure gas.[16,17] A known volume V_L of the analyte solution is collected with a glass syringe under thermostatic conditions (a container with variable volume) together with gas of volume V_G. After equilibrium is established, the initial concentration of the substance in the gas phase (C_G) is determined gas chromatographically. Then the gas phase is substituted with an equal volume of pure gas. Once equilibrium is again established, the concentration C'_G is determined. With these data, determined in an interval of concentrations from C_L to C'_L, the isotherm of the distribution is linear:

$$K = \frac{C_L}{C_G} = \frac{C'_L}{C'_G}$$

The value of the distribution coefficient can be calculated from the expression

$$K = \frac{C'_G}{C_G - C'_G} \frac{V_G}{V_L} \tag{1.22}$$

Within the linear range of the chromatographic detector, and under constant analytical conditions, the peak areas (or heights) A_G and A'_G, which are proportional to the substance determined can be substituted for the values of C_G and C'_G respectively:

$$K = \frac{A'_G}{A_G - A'_G} \frac{V_G}{V_L} \tag{1.23}$$

Equation (1.23) increases the accuracy of measurements because the distribution coefficient is calculated without determining the absolute values of the substance's concentration in the equilibrium gas phase.

One can evaluate the boundaries of the application of the method (by numerical values of K) in the following manner. The lower limit, the minimum value of the distribution coefficient K^{min}, can be obtained from Equation (1.22) if one assigns the assumed decrease of concentration of the analyzed substance $(C'_G/C_G)^{lim}$ when replacing the equilibrium gas by pure gas. The value $(C'_G/C_G)^{lim}$ depends on the initial concentration of the substance and on the relationship between the volumes of the phases. When evaluating the limiting value of this relationship, it is necessary to begin with the sensitivity of the detector, the limitations of which usually are known. In reality, if C_G is registered by the detector near the limit of detection, then C'_G can either be determined with a large error margin or will not be determined, since the height of the peak will be unmeasurable or will occur within the background noise. However, even in that case, if the detector has a high sensitivity, the value of $(C'_G/C_G)^{lim}$ should not be below 0.01, since if $C_G \gg C'_G$ the error in the determination of C'_G (or A'_G) increases sharply. For instance, if $(C_G/C_G)^{lim} = 0.05$ and $V_G = V_L - K^{min} = 0.06$, when the experimental conditions allow V_G/V_L to become 0.1, K^{min} will decrease to 6×10^{-3}.

In practical applications the given values of K^{min} do not limit the variety of substances where this method can be applied. Therefore, when considering the possibility of the determination of the distribution coefficient of a substance, it is only necessary to consider the maximum limit of the distribution coefficient K^{max}. These values can be obtained by analyzing the dependence of the error in the determination of the distribution coefficient from its value and its relationship to the volumes of the phases.

From Equation (1.22), after differentiation we have

$$\frac{\Delta K}{K} = \left(\frac{\Delta C_G}{C_G} + \frac{\Delta C'_G}{C'_G}\right)\left(\frac{C_G}{C_G - C'_G}\right) = \left(\frac{\Delta C_G}{C_G} + \frac{\Delta C'_G}{C'_G}\right)\left(\frac{KV_L}{V_G} + 1\right) \quad (1.24)$$

This expression indicates that the error in the determination of K decreases with the increase of V_G/V_L as well as with the decrease of the numerical value of the distribution coefficient. Values of K less than 10 can be determined with an accuracy up to 3% if the total error in the determination of the concentrations C_G and C'_G is 2% and if $V_G/V_L = 20$. If a 5% error is allowed in the determination, then K^{max} increases to 30.

To increase the upper limit of the determination of K, the number of substitutions of the equilibrium gas by pure gas must be increased. Thus, in accordance with the elementary theory of extraction, after the nth substitution the concentration of the substance in the equilibrium gas, C_G^i, is related to the initial concentration C_G

$$C_G^i = C_G \left(\frac{K}{K + V_G/V_L} \right)^n \qquad (1.25)$$

where

$$K = \frac{(C_G^i)^{1/n}}{(C_G)^{1/n} - (C_G^i)^{1/n}} \frac{V_G}{V_L} \qquad (1.26)$$

From Equation (1.25), it follows that

$$\frac{\Delta K}{K} = \frac{1}{n} \left(\frac{\Delta C_G}{C_G} + \frac{\Delta C_G^i}{C_G^i} \right) \left(\frac{KV_L}{V_G} + 1 \right) \qquad (1.27)$$

Therefore, increasing the number of substitutions from 1 to n decreases the error in the determination of K and increases K^{max} by n times.

McAuliffe[16] first described this method concerning the determination of the distribution coefficient and recommended a graphical method for finding K using the value of the angle coefficient (Ψ) of the dependence of $\ln C_G^i(n)$ (in the system accepted by us), that is, according to Equation (1.25).

$$\Psi = \ln \left(\frac{K}{K + V_G/V_L} \right) \qquad (1.28)$$

This method of determining K with several extractions requires the measurement of the intervening concentrations of the substance in the gas phase (or peak areas in the chromatogram) between C_G and C_G^i. These measurements considerably increase the time and work involved in the analysis, but with a large number of measurements, the accuracy of K as determined by this method will increase to a certain extent. An advantage of the graphical method is the simultaneous check of the constancy of K while completing the analysis in a studied concentration range. This is possible since Ψ remains unchanged in this case. Therefore, regardless of the work involved, the graphical method is preferred in cases in which the distribution isotherm in a studied concentrations range has not been proven to be linear.

Equation (1.26) is similar to Equations (1.22) and (1.23); The values of the concentrations are proportional to the peak areas or heights in the

1.2 MEASURING THE DISTRIBUTION COEFFICIENTS

chromatogram; therefore, the latters can be substituted for concentrations. The computation of K is completed without determining the absolute values of the concentrations.

The method of substituting pure gas for the equilibrium gas was utilized in references 16–18 for the determination of the distribution coefficients of paraffins and aromatic hydrocarbons and of the simplest sulfur-mercury and organomercury compounds between water and air. In these cases the values of the distribution coefficients are low ($K < 20$). The reproducibility of the determination is 3–8%. The discrepancies between the K values obtained by this method and published values measured by other methods do not exceed the reproducibility of this method.

The second version, which is preferred for the determination of medium values of K (approximately in the range of 50 to 10^3), consists of introducing a solvent volume V_L into a fixed volume of gas V. The gas contains the vapors of the substance of concentration C_G^0 distributed between the liquid and gas. After equilibrium is established, the concentration of the liquid and gas phases is given by C_L and C_G, respectively.

This case was developed in containers with constant[13,19] and varying[17,20] volume. Fixed-volume systems are characterized by securing a constant pressure in the system when the introduction of the solvent is accompanied by the displacement of the gas containing the distributed substance. This decreases the initial mass (VC_G^0) to the extent of $V_L C_G^0$. The volume of the gas phase in the system becomes $V_G = V - V_L$. The equation of mass balance can be written in the form

$$VC_G^0 - V_L C_G^0 = V_G C_G + V_L C_L \qquad (1.29)$$

from which, considering the value of the distribution coefficient as described by Equation (1.3), one obtains the working formula

$$K = \frac{C_G^0 - C_G}{C_G} \frac{V_G}{V_L} \qquad (1.30)$$

which relates the distribution coefficient to the concentration of the substance in the gas phase before and after the introduction of a solvent into the system.

The shortcoming of fixed-volume applications lies in the introduction of a large volume of liquid. This results in a noninstantaneous displacement of the gas from the system with the accompanying extraction of a certain amount of the substance. Thus its concentration in the displaced gas will be lower than C_G^0. To take into consideration the quantity displaced by the liquid upon its introduction into the container is extremely complicated. Therefore, the use of fixed-volume instrumentation is restricted to high values of the ratio V/V_L, when the fraction of the displaced substance can be disregarded. This method for determining K under such conditions and assumptions for constant-volume containers is discussed by Nelson and Hoff[13] for the measurement of the distribution coefficients for a series of alcohols, carbonyl compounds, and complex ethers in water and water–salt solutions. Hasty[19] used the method to obtain data on the solubility of methyl iodide. In the latter work, the calculation of K was completed by an approximation equation:

$$K = \frac{V_G(C_G^0 - C_G) + V_L C_G^0}{V_L C_G} \tag{1.31}$$

The equation is obtained from (1.29) for $V_L C_G^0 \ll V_G C_G^0$.

The need to use large volumes of gas and small quantities of liquid in the constant-volume system considerably limits the variety of substances, especially with low values of K, for which sufficient accuracy can be obtained. From Equation (1.30) it follows that

$$\frac{\Delta K}{K} = \left(\frac{\Delta C_G^0}{C_G^0} + \frac{\Delta C_G}{C_G}\right)\left(\frac{V_G}{KV_L} + 1\right) \tag{1.32}$$

This expression allows the evaluation of K^{min} under three conditions: a given error of K, a known total error in the determination of the concentration, and a given relationship between phase volumes. For example, if the error in the determination of K must be less than 5% and the total error in the determination of the concentration in the gas is 2%, at $V_G/V_L = 100$ (these conditions are close to those used by the authors of this method[13]), then $K^{min} = 66$ and for $V_G/V_L = 500$, $K^{min} = 330$.

The values of K^{max} depend upon the accuracy with which low residual concentrations C_G in the gas can be determined after the introduction of the liquid into a vapor–gas mixture. If the accuracy of measuring the ratio C_G/C_G^0 can be held to several percentage points while decreasing this relation to ~ 0.01, then in the case where $V_G/V_L = 100$, according to Equation (1.30), $K^{max} = 10^4$, but at $V_G/V_L = 10$, K^{max} decreases to $\sim 10^3$.

The discussed limitation of the method caused by the use of constant-volume apparatus can be excluded if variable-volume containers are used to achieve the equilibrium distribution of the substance between the liquid and

1.2 MEASURING THE DISTRIBUTION COEFFICIENTS

a gas phases. These containers should be constructed of low-sorbency materials.[17] Such containers include glass syringes with a volume up to 100 ml (see Chapter 2). In this case, while introducing the solvent into a container of increasing volume, the reduction or displacement of the gas does not occur and the computation of K is completed with the use of Equation (1.30). The use of apparatus with varying volume removes the limitations regarding the quantity of solvent that can be introduced into the vapor–air mixture. This fact considerably widens the range of the distribution coefficients that can be determined by this method. Thus, reducing the ratio V_G/V_L to ~ 1 allows the lowering of K^{min} to values near 1.

A considerable contribution to the error in the determination of K of polar substances occurs due to the adsorption of the substance onto the walls of the container. These losses can be measured by using the method[17] based upon the establishment of the change in the concentration of the substance in a volume of gas (or liquid) while introducing it into a vessel. The adsorption activity of the walls of the container is simply determined with the use of two syringes connected by a double-edged needle. Syringe I is filled with the vapor–gas mixture with a concentration of the analyzed substance C_G^I. The concentration is determined by the area A_G^I of the peak in the chromatogram. Then gas from syringe I is pumped into syringe II, such that the concentration of the substance acquires the value C_G^{II}. Under similar conditions the concentration is determined by the peak area A_G^{II}. If both C_G^I and the volume of the transferred gas V_G are known, the quantity Q of the substance adsorbed can be determined from the expression

$$Q = C_G^I V_G \left(\frac{A_G^I - A_G^{II}}{A_G^I} \right) \qquad (1.33)$$

Once the quantity of the substance adsorbed has been determined, the distribution coefficient can be computed from the equation

$$K = \frac{V_G(C_G^0 - C_G) + Q}{V_L C_G} \qquad (1.34)$$

The operations of determining Q and measuring K can be combined. In this case, gas from syringe I is forced into syringe II, into which a solvent is also introduced. With these conditions C_G^0 becomes C_G', and the computation is carried out using Equation (1.30).

The use of Equation (1.34) assumes that the adsorption of the substance by the walls from a solution is much less than that from the gas phase. If this is not a realistic assumption, it is not difficult to evaluate the adsorption of the substance from the solution by using the above-described method. The value of Q represents the difference between the adsorption of the substance on the dry walls of the container and that upon the solvent-wetted walls.

By the introduction of a solvent into the vapor–gas mixture, values of K for benzene and toluene in acetic acid and of acetone and methyl ethyl ketone in water[17] were measured. These values differed from the literature values by not more than 6%.

The third version is based on an exponential decrease in the concentration of the volatile components present in a solution while a flow of pure gas is passed through it. This case is preferred for the determination of medium and large values of K (>50–1000).

The process of the gradual extraction of the volatile components from a solution using a flow of pure gas (continuous gas extraction) can be described by the following relations.[21,22]

During the passage of a microbubble of gas or volume dv_G through V_L ml of solution and the establishment of equilibrium, in accordance with the value of the distribution coefficient given by Equation (1.3), the decrease in the content of the volatile substance in the solution of initial concentration C_L^0 will be equal to its quantity in the volume of the bubble, that is,

$$-V_L \, dC_L = \frac{C_L}{K} \, dv_G \qquad (1.35)$$

After separating the variables and integrating over the interval from C_L^0 to C_L and from 0 to v_G, one obtains

$$\ln \frac{C_L^0}{C_L} = \frac{v_G}{KV_L} \qquad (1.36)$$

It is assumed that the effects related to the curvature of surface of the bubbles, hydrostatic pressure, evaporation of solvent, and dissolved gas can be disregarded. In the case of volatile solvents considerably dissolving the blown-through gas, the effect of these factors can be excluded by first saturating the gas with vapors of pure solvent and then saturating the solution with this gas.

The process of continuous gas extraction in practical applications is accomplished in a container of a known gas volume V_G. Equation (1.36) does not take into consideration the effect of this volume on regularities in the changes of the concentrations of the substances in solution and in the gas

1.2 MEASURING THE DISTRIBUTION COEFFICIENTS

flow at the outlet of the container. Therefore, it is applicable only under the condition in which $KV_L \gg V_G$ (for details, see Section 5.3).

Measuring the distribution coefficient consists of determining the change in the concentration of a solution or gas that has passed through this solution. As such, this concentration is a function of the volumes of the bubbled gas. If the liquid phase is analyzed in order to determine K, the values C_L^0 and C_L are measured as peak areas (or heights) A_L^0 and A_L in the chromatogram. The value is obtained from the equation

$$K = \frac{v_G}{V_L \ln(A_L^0/A_L)} \quad (1.37)$$

When a gas that has passed through a solution of a substance is subjected to analysis, the computation of K requires measurement of the values A_G and A'_G that are proportional to the concentrations C_G and C'_G for two different volumes of gas v_G and v'_G ($v_G < v'_G$). In this case the calculation is made using the equation

$$K = \frac{v'_G - v_G}{V_L \ln(A_G/A'_G)} \quad (1.38)$$

Conditions regarding an exact measurement of K must be carefully selected as to the volume of the gas used (v'_G) and the conditions of achieving equilibrium. Passing an insufficient volume of gas through the solution will lead to insufficient changes in the concentration of the substance in it and large errors in the determination of K. From Equation (1.37), disregarding the error in the measurement of the liquid and gas volumes, it follows that

$$\frac{\Delta K}{K} = \left(\frac{\Delta A_L^0}{A_L^0} + \frac{\Delta A_L}{A_L}\right) \frac{1}{\ln(A_L^0/A_L)} \quad (1.39)$$
$$= \left(\frac{\Delta A_L^0}{A_L^0} + \frac{\Delta A_L}{A_L}\right) \frac{KV_L}{v_G}$$

The ratio A_L^0/A_L can serve as a criterion to determine the quantity of gas to be admitted. This ratio must not be smaller than the number e (i.e., it must be ≥ 2.7). In this case, the error in the determination of K will not exceed the total relative error of A_L^0 and A_L. In using Equation (1.38) for the calculation of K, the difference between the volumes, $v'_G - v_G$, must meet this condition.

Practically, an equilibrium process of continuous gas extraction can be achieved e.g. when using water solutions, by introducing the gas into the solution through a bubbling apparatus.[22] The instrumentation consists of 5–

36 THEORY OF GAS-CHROMATOGRAPHIC HEAD-SPACE ANALYSIS

7 steel capillary tubes or a fine-pore glass filter with a velocity of up to 200 ml/min. A major advantage of this method is the possibility of measuring high values of K (up to 10^5). The determination of such values faces serious difficulties when static methods are used. In reality, it follows from Equation (1.39) that values of K^{max} are limited by the volume of gas passing through the solution. If, under equilibrium conditions, 10 liters of gas can be passed through 1 ml of liquid such that the error in the determination of K is permissible (5%), then the total error in the peak area is $0.01 - K^{max} = 2.5 \times 10^4$.

The conformity of the basic relations (1.37) and (1.38) is illustrated in Figure 1.2 by an example of diluted solutions of the simplest alcohols and benzene in ethylene glycol and polyethylene glycol 300.[22] The slope of these curves is determined by the magnitude of the distribution coefficient. The linearity of the curve characterizes the stability of K in a studied interval of concentration.

Using one of these versions selected according to magnitude of the

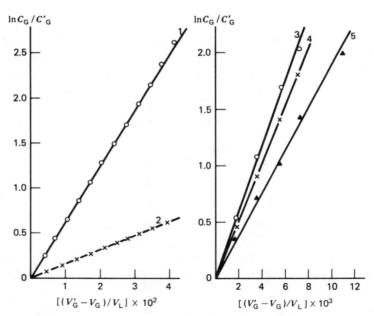

Figure 1.2. Concentration change in dilute solutions and in their vapors under continuous extraction with nitrogen. *1, 2*, Benzene in ethylene glycol and polyethylene glycol 300 at 20°C; *3, 4, 5* n-butanol, n-propanol, and ethanol, respectively, in water at 25°C. Initial concentration (C_L^0) of benzene and ethanol is 1%, and of n-propanol and n-butanol is 0.1%.

1.3 STATIC METHODS OF HEAD-SPACE ANALYSIS

distribution coefficients, K can be determined over a wide range of values (from hundredths of a unit to 10^5 units). The method chosen depends upon the properties of the analyzed substance and the allowed error of determination.

1.3 STATIC METHODS OF HEAD-SPACE ANALYSIS

Single-stage gas extraction of a substance from a solution

The procedure for this, the simplest case of head-space analysis, consists of the following aspects (Fig. 1.3). The investigated solution, with volume V_L and concentration of the analyzed substance C_L^0, is placed into a fixed-capacity container of volume V. The container is held at a constant temperature until the equilibrium for the substance is established between the liquid and gas phases. In the process of equilibration the analyzed substance in the solution partially changes into the gas phase. At equilibrium, its concentrations in the liquid and gas are C_L and C_G, respectively. In this way the simplest case of the HSA method can be considered a one-stage extraction by the vaporization of the analyzed substance from a solution.

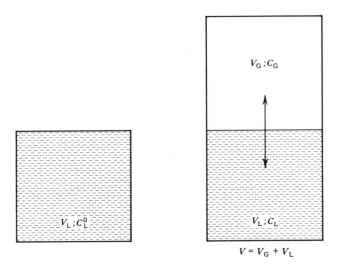

Figure 1.3. Schematic of a single extraction of a substance from a solution using a gas. See explanation in the text.

The quantity of the substance extracted from the solution by means of the gas phase depends on the relationship between the volumes of the phases and the value of the distribution coefficient. If the change in the volume of the liquid due to the evaporation of solvent during equilibration is disregarded, the concentration of the substance in the initial solution can be computed from its amount in the equilibrium gas using Equation (1.20).

The basic parameters determining the sensitivity of an analysis are the values of the distribution coefficient and the relationship of the phase volumes in the container in the establishment of equilibrium. In reality, the sensitivity S of the HSA method is expressed by the ratio

$$S = \frac{A_G}{C_L^0} \qquad (1.40)$$

For the area of the peak on a chromatogram, we can write

$$A_G = fm_g = fv_g C_G \qquad (1.41)$$

where f is a coefficient expressing the sensitivity of the chromatographic detector for the substance of interest, m_g is the mass of the substance introduced into the chromatographic column per sample of equilibrium gas, and v_g is the volume of sample of the equilibrium gas introduced into the column.

Substituting into (1.40) the values of A_G from (1.41) and C_L^0 from (1.20), one obtains the equation

$$S = f \frac{v_g}{K + V_G/V_L} \qquad (1.42)$$

which establishes the relationship of the sensitivity of head-space analysis with the nature of the analyzed substance, the liquid phase, and the conditions under which the experiment is conducted.

The sensitivity of HSA methods can vary due to the fact that the values of K of the volatile substances in various systems and conditions can vary over an extremely wide range (from hundredths to 10^6). Therefore, first to be discussed is the expediency of utilizing head-space analysis to increase the sensitivity of the determination of a substance in a solution. This method will be compared to the direct introduction of the investigated sample into a chromatographic column (which has great value during the analysis of impurities).

It is apparent that an increase in HSA sensitivity is obtained when the

1.3 STATIC METHODS OF HEAD-SPACE ANALYSIS

amount introduced into the chromatograph with the head-space sample is higher than the amount of the direct introduction of a liquid sample, that is, $C_G v_g > C_L^0 v_L$ (v_L is the volume of a single sample of the liquid introduced into the chromatographic column). This also means that the following relationship (α), representing the ratio of the quantities of a substance introduced into the chromatograph, is a measure of the change in the sensitivity of head-space analysis relative to the direct introduction of the investigated solution

$$\alpha = \frac{v_g C_G}{v_L C_L^0} = \frac{v_g}{v_L(K + V_G/V_L)} \qquad (1.43)$$

Since the limiting volumes of the liquid introduced into the chromatograph are on the order of several microliters, and the limiting volume of the gas is several milliliters, $v_g/v_L \simeq 10^3$, and Equation 1.43 can be written as

$$\alpha = \frac{10^3}{K + V_G/V_L}$$

This relationship indicates that the gain in sensitivity of the analysis ($\alpha > 1$) is reached when $K < (10^3 - V_G/V_L)$, that is, with average and especially low (<10–30) values of K. If the limit of detection of a dissolved substance is considered, the direct introduction of the solution into a gas chromatographic column employing a universal flame-ionization detector allows a limit of usually 10^{-4}–10^{-5}% (1–0.1 ppm). Therefore, if $K = 10$ the analysis of the equilibrium vapor above the solution to be analyzed allows a reduction of this value to 10^{-6}–10^{-7}% (10–1 ppb). At values of $K > 10^3$, the sensitivity of the analysis decreases, and the use of head-space analysis must be dictated by other considerations. Such factors include the impossibility of the direct introduction of a solution, possible contamination of the column by a nonvolatile substance, etc.

Another factor affecting the sensitivity of head-space analysis is the relationship between the volumes of the phases. This parameter considerably influences the sensitivity of the analysis at low values of K. Therefore correct selection of the parameters concerning the volumes of the phases is an important condition of successful and possibly more sensitive and more accurate analysis.

Equation (1.42) allows an increase in sensitivity at a given V_L if V_G is made as small as possible while allowing for the necessary number of determinations. However, reducing the ratio V_G/V_L to a minimum value somewhat increases the error in the determination of the initial concentration of the substance in the solution, $\Delta C_L^0/C_L^0$. From Equation (1.20) it follows that

$$\frac{\Delta C_L^0}{C_L^0} = \frac{\Delta C_G}{C_G} + \frac{\Delta K}{K} \frac{K}{K + V_G/V_L} \qquad (1.44)$$

This equation indicates that the increase of the ratio V_G/V_L lowers the contribution of the error in the determination of K to the total error of an analysis (especially in the most favorable cases with low K). Considering that $\Delta K/K$ usually exceeds $\Delta C_G/C_G$ if the conditions of gas-chromatographic analysis ensure sufficient sensitivity of the determination of C_G, the ratio of V_G/V_L should be increased to the limits allowed by the system used to establish the equilibrium distribution of the substance between the phases.

Traditional quantitative gas-chromatographic techniques (absolute calibration and addition of an internal standard to the investigated solution) can be used to determine the initial concentration of a substance in solution from its concentration in the gas phase.

There are two main variations for absolute calibration. The first correlates the peak area or heights on the chromatogram with the concentration of the substance in the analyzed solution, A_G vs. C_L^0. In the other, the peak area is correlated to the concentration of the substance in the equilibrium gas phase (C_G).

Calibration of A_G vs. C_L^0. This is accomplished using prepared standard solutions of the substance with known concentrations. These standards in an equilibrium vessel have a constant, but not necessarily exactly known, ratio of the volumes of the phases. The values of V_G/V_L in the vessels containing the standard and investigated solutions and the volumes of the samples introduced into the chromatograph should be equal. For single measurements it is sufficient to prepare one standard solution with the concentration C_L^{st}, for which the peak area on the chromatogram A_G^{st} is covariant with the peak area of the compound being determined in the analyzed sample A_G. The sought concentration can be computed from the simple relationship

$$C_L^0 = C_L^{st} \frac{A_G}{A_G^{st}} \qquad (1.45)$$

If the analysis carried out for a larger number of samples over a wide range of concentrations or in a nonlinear area of the distribution isotherm, the

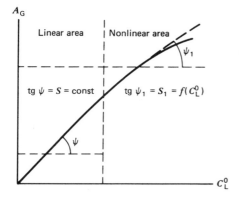

Figure 1.4. Relationship of the peak area in a chromatogram to the concentration of a substance in the initial solution. ψ and ψ_1 are the angles of the slopes for the linear and nonlinear areas of the isotherm.

computation of C_L^0 is preferably completed graphically according to a previously constructed calibration plot A_G vs. C_L^0 (Fig. 1.4). One of the advantages of the graphical method is the possibility of conducting quantitative head-space analysis in cases where K depends upon C_L. This simple method, which has acquired wide recognition in analytical practice, is still suffering from certain limitations that must be considered if gross oversights and large errors are to be avoided.

1. The characteristic of this method of calibration, in which the peak area on the chromatogram is correlated to the concentration of the substance in a solution, lies in the computation in which it is unnecessary to know the absolute value of K. This does not mean that the quantitative analysis is completed upon a system with a known value of K. In conducting the calibration A_G vs. C_L^0, the value of K is inherently determined because the slope of the plot (Fig. 1.4) represents the sensitivity of head-space analysis, which is a function of K. In other words, this method of calibration causes the distribution coefficient to appear as part of the total coefficient of calibration. This coefficient takes into consideration not only the value of K, but also the effects of the volume of the phases and the sensitivity of the detector. Therefore, the properties of the standard solutions and the analytical conditions must be adequate to allow acceptable results in the investigated samples. If such is not the case, the change of K can lead to additional errors as well as errors that may be difficult to detect.

2. By the same reasoning the calibration of A_G vs. C_L^0 can be used only for those substances with constant and unchanging solution properties for different samples.

3. In the preparation of calibration solutions containing highly volatile substances (with low K), consideration must be given to the possibility of a change in the given value of C_L^{st}. This change is due not only to the transition of a portion of the volatile substance into the gas phase in the container, but also to its evaporation in the process of filling or obtaining a sample. For instance, if a solution of benzene in water is being prepared at a temperature of 20°C ($K = 5$) and the container is being half-filled with the solution ($V_G/V_L = 1$), then Equation (1.20) will have the form

$$C_L = C_L^0 \frac{K}{K + V_G/V_L}$$

It is easy to calculate that the transition of the substance into the gas space of the container causes the benzene concentration in the solution to be 17% less than that given. The difference in effect between large values of K and low values of V_G/V_L has a leveling action. Therefore, in order to avoid this type of error, calibration solutions are to be prepared in containers filled to capacity. Samples of the standard solution are then taken through a rubber septum using a syringe whenever needed.

Calibration A_G vs. C_G. A method of absolute calibration is the standardization of the chromatograph referenced against the concentration of the substance in the gas phase. The slope of the A_G vs. C_G plot differing from that in the above-described method considers only the sensitivity of the detector with respect to a given substance (Fig. 1.5). Therefore, in accordance with Equation (1.20) the absolute values of the distribution coefficient and the phase-volume relationship in the equilibration container are suffi-

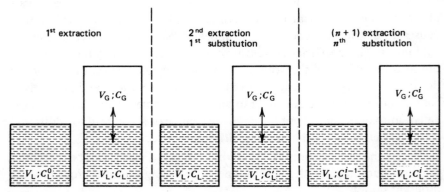

Figure 1.5. Schematic of multiple gas extraction of a substance from a solution. For the symbols, see the text.

cient for the calculation of the initial concentration of a substance in a solution.

If the substances analyzed have constant values of the distribution coefficient in different samples, the values of K can be measured initially by one of the above-described methods (Section 1.2). These values can then be used in calculating the analyses for an entire series. If the substance in the analysis forms solutions with varying properties (varying values of K are caused by changing the composition of a solution or by the distribution isotherm being nonlinear), the analysis must include the determination of K characteristic of the given sample.

Measuring the A_G vs. C_G relationship requires the preparation of a series of gas mixtures over the concentration range comparable to those observed in the gas in equilibrium with the analyzed solution. Known methods including diffusion can be used to this end. The methods of standardizing the chromatographic detectors based on the gas extraction of the substance from a solution (discussed in detail in Section 5.3) are quite promising.

Gas mixtures prepared by one of the foregoing methods are introduced into the chromatograph. The sample size can be unknown but it must be constant, both during the calibration and during the analysis of the investigated samples. The best results are obtained when gas samples are injected through thermostated gas sampling valves (see Chapter 2).

Internal standard method. The procedure for completing head-space analysis by this method is similar to its traditional use in chromatography. A standard substance of concentration C_L^{st} is added to the investigated solution. This system, in agreement with (1.20), defines the equilibrium concentration of the standard in the gas phase C_G^{st} by the relationship

$$C_L^{st} = C_G^{st} \left(K^{st} + \frac{V_G}{V_L} \right) \tag{1.46}$$

The equilibrium gas is introduced into the chromatograph and the concentrations C_G^{st} and C_G are measured as the areas A_G^{st} and A_G of peaks on the chromatogram.

The derived equation for the determination of the substance in the investigated solution, using HSA techniques and an internal standard, relates the initial concentration of the substance and the internal standard in the solution to their equilibrium concentrations in the gas phase:

$$\frac{C_L^0}{C_L^{st}} = \frac{C_G}{C_G^{st}} \frac{K + V_G/V_L}{K^{st} + V_G/V_L} \tag{1.47}$$

In the limits of a linear range of the chromatographic detector,

$$\frac{C_G}{C_G^{st}} = f_r \frac{A_G}{A_G^{st}}$$

where f_r is the coefficient that takes into consideration the relative sensitivity of the chromatographic detector toward the analyzed and standard substance, and Equation (1.47) will finally acquire the form

$$C_L^0 = f_r C_L^{st} \frac{A_G}{A_G^{st}} \frac{K + V_G/V_L}{K^{st} + V_G/V_L} \tag{1.48}$$

In current HSA methods using an internal standard, the following equation is used for calculations:

$$C_L^0 = k C_L^{st} A_G/A_G^{st} \tag{1.49}$$

where k is a calibrating factor that is usually experimentally determined.

Comparing Equations (1.48) and (1.49), it is obvious that the numerical value of the calibration factor depends upon the parameters of the analyzed system and upon the relative sensitivity of the chromatographic detector to the determined substance and the standard:

$$k = f_r \frac{K + V_G/V_L}{K^{st} + V_G/V_L} \tag{1.50}$$

As the values of K and K^{st} approach equality, or when the ratio of the phase volumes (V_G/V_L) becomes much greater than K and K^{st}, the calibration factor becomes the coefficient of the relative sensitivity of the chromatographic detector. However, if such conditions are not fulfilled, then k depends upon the values of the distribution coefficients of a standard and a given substance and must be obtained experimentally. To this end, the gas over the standard solution with known C_L^0 and C_L^{st} (at a fixed temperature and V_G/V_L ratio) is analyzed. Equation (1.49) is used to complete the calculations. It must be noted that the properties of the standard solution (values K and K^{st}) must be identical to those of the investigated samples. Therefore, the analysis of substances with changing solution properties in various samples using an internal standard can lead to considerable error. Although these concepts should be self-evident, in analytical practice there are cases when the peculiarities of the use of an internal standard related to the possibilities of changes in the distribution coefficients in various samples are not taken into consideration.

1.3 STATIC METHODS OF HEAD-SPACE ANALYSIS

The advantages of head-space analysis with an internal standard include good reproducibility and high accuracy in the measurement of the ratio of areas (or heights) of the peaks. As such, the method does not require exactly reproducible sample volumes of the equilibrium gas. This simplifies the instrumental design of head-space analysis and allows the use of common as well as special syringes. Although the use of an internal standard is expedient when determining the impurities in solutions (especially with low values of K), one must consider the complications and the work necessary in preparing solutions of volatile substances with concentrations of ppm and less. Therefore, applications using an internal standard can be expedient for single measurements. However, for series analyses, the use of absolute calibration must be considered to be more justified. This is also advisable because modern instrumentation allows great accuracy in the setting and reproducibility of analytical conditions.

As a conclusion we may state that the simplest version of the HSA methods is based on a single gas extraction of a substance from a solution. This method requires that the distribution coefficient of the investigated substance in various samples to be analyzed is known in advance either in numerical value or in factor form. Frequently, the composition of the investigated materials can fluctuate over wide ranges, so that the relationship of K to the content due to other components cannot be ignored. Under these conditions, it is impossible to consider the constant K as identical for all investigated samples. For instance, considerable fluctuations in the content of mineral salts in natural waters, which are analyzed for trace hydrocarbons, noticeably affect the distribution coefficients of these substances. When determining the concentrations of volatile organic substances in industrial waste and in biological matter, similar complications arise in relation to the fluctuations of the total quantity of dissolved impurities. In such cases one must utilize the variations of the HSA method that require no prior knowledge of the numerical values of K but include their determination as a procedure of analysis.

Multiple gas extraction

The method of impurity determination in a solution under static conditions of equilibrium was first suggested by McAuliffe[16] and later modified in the laboratory of gas chromatography of Leningrad University[23,24]. The case in which the equilibrium distribution of the substance between the phases is achieved does not require a previous knowledge of the value of K for the analyzed substances. This version of head-space analysis consists of the gas-chromatographic determination of the change in concentration of the analyzed substance in the equilibrium gas phase of the investigated solution. This is achieved by its successive substitution by an equal volume of pure air (or nitrogen).

The analysis includes the following operations (Fig. 1.5). First, the investigated solution, the same as described in the simplest version of head-space analysis, is placed into a known volume (usually a glass syringe having a volume of 50–100 ml) and is brought to equilibrium with a gas that does not contain the substance being determined. After equilibration the gas-chromatographic determination of C_G is conducted by completely substituting equilibrium gas for pure air (or nitrogen). The remaining quantity of the substance in solution ($V_L C_L$) redistributes between the two phases. This substitution of equilibrium gas by a pure gas is repeated n times. The first extraction is used to determine the absolute value of C_G, and subsequent extractions are used only for the measurement of K (see Section 1.2). Therefore, the formula for the determination of the volatile components of a solution utilizing multiple extractions under static conditions can be derived by substituting the value of K from Equation (1.26) into Equation (1.20).

$$C_L^0 = C_G \frac{(C_G)^{1/n}}{(C_G)^{1/n} - (C_G^i)^{1/n}} \frac{V_G}{V_L} \tag{1.51}$$

Considering the characteristics of gas-chromatographic determinations using HSA methods, Equation (1.51) can be modified. This is done by taking into consideration that the relation $C_G/(C_G - C_G^i)$ at equal volumes of gas introduced into the chromatograph is equal to the similar function of the corresponding parameters of the peaks A_G and A_G^i (areas or heights). In such a case,

$$C_L^0 = C_G \frac{(A_G)^{1/n}}{(A_G)^{1/n} - (A_G^i)^{1/n}} \frac{V_G}{V_L} \tag{1.52}$$

The value C_G is found with the help of a calibration graph C_G vs. A_G, which, unlike other factors, includes the errors in the calibration of the chromatograph. The ratio of the phase volumes V_G/V_L is measured comparatively easily and accurately, and the error of its measurement can be disregarded.

1.3 STATIC METHODS OF HEAD-SPACE ANALYSIS

Thus one can, in the case of multiple gas extraction, obtain the error of the analysis of interest by substituting the value $(\Delta K)/K$ from Equation (1.27) into (1.44):

$$\frac{\Delta C_G^0}{C_L^0} = \frac{\Delta C_G}{C_G} + \left(\frac{\Delta A_G}{A_G} + \frac{\Delta A_G^i}{A_G^i}\right)\frac{KV_L}{nV_G} \quad (1.53)$$

This relationship demonstrates the analytical possibilities of this HSA method. At a given accuracy of chromatographic measurements, the errors in the determination of the concentration C_L^0 increase linearly with the increase of the factor $KV_L/(nV_G)$. Therefore, the basic factors of an experiment, which must be selected to achieve better accuracy with the least amount of work, are the necessary number of substitutions n and the ratio of the volumes of the phases V_G/V_L in a container for establishing equilibrium. For instance, the relative error of determination of the initial concentration of a substance in a solution will not exceed the total error in the measurements of C_G, A_G, and A_G^i of $K \leqslant nv_G/V_L$. For this reason, the high values of K of the analyzed substances must be compensated by increasing n and V_G/V_L. However, completing a large number of substitutions demands a considerable increase in the work involved in the analysis. Therefore, it is questionable whether it is beneficial to carry out more than four to six substitutions (the time needed for one substitution is 20–30 min). It is inconvenient (and impossible) to considerably increase the ratio V_G/V_L in fixed-volume containers. Thus, it is recommended that for conducting such analyses, devices with varying volumes be used (see Chapter 2) such that the value of V_G/V_L does not exceed 10–30. From this fact it is evident that the possibilities of this version of head-space analysis are limited to the determination of volatile components having average (on the order of hundreds) and low values of the distribution coefficients.

Method of addition of the substance being determined to an investigated sample

This variation of the HSA method, which allows the analysis of systems with unknown distribution coefficients, was first discussed in Novák's monograph[25] and confirmed experimentally by him.[26]

The technique of conducting the analysis is in many ways similar to that using an internal standard. It differs in that the area of the peak on the chromatogram is determined twice (before and after making the addition). The mass of the analyzed substance in the initial solution, obeying Equation (1.20), can be written as

$$m_0 = V_L C_L^0 = C_G(KV_L + V_G) \tag{1.54}$$

After a sample is selected for analysis (v_g with concentration C_G, that is, mass $m_v = v_g C_G$), and a mass m_s of the analyzed substance is added to this system, the equilibrium concentration of a substance will change in the gas phase to the value C_G'. For this altered state of the system, one can write

$$m_0 - m_v + m_s = C_G'(KV_L + V_G) \tag{1.55}$$

Equating formulas (1.54) and (1.55) gives the expression

$$C_L^0 = \frac{m_s - m_v}{V_L} \frac{C_G}{C_G' - C_G} \tag{1.56}$$

If the values of peak area, which are proportional to concentration, are substituted into this equation, one obtains

$$C_L^0 = \frac{m_s - m_v}{V_L} \frac{A_G}{A_G' - A_G} \tag{1.57}$$

Equation (1.57) allows the determination of the initial concentration of a substance in solution, without having data on the absolute values of the distribution coefficient. Besides the ratio of the peak areas on the chromatogram A_G'/A_G and the investigated solution volume, it is necessary that the masses selected for analysis and introduced into the system be measured accurately. This can be achieved easily from the value of A_G, with help of previous calibration of A_G vs. m_v.

The important condition for an accurate analysis is the correct selection of the relationship between the initial, introduced and withdrawn quantities of the substance selected for the analysis. The dependence of the error of the analysis upon experimental conditions indicates that the mass of the substance added (m_s) must considerably exceed (at least by several times) the initial quantity of the substance (m_0) and the quantity withdrawn from the system (m_v) in order to obtain acceptable accuracy.

The advantage of the method as compared to the HSA variations based on gas extraction lies in the unlimited numerical values of the distribution coefficients of the analyzed substances, with the detection range remaining the same as in the simplest method (single extraction). However, the use of this method requires a great deal of work, since it includes the operations that are being carried out both in the absolute calibration and in the introduction of an internal standard. Besides this, if the method of absolute calibration requires only the constancy of the sample volume of gas, then in the discussed case it is necessary to know the absolute size of the introduced sample. This can be avoided if the condition $m_v \ll m_s$ is imposed.

1.4 DYNAMIC VARIATIONS OF HEAD-SPACE ANALYSIS—CONTINUOUS GAS EXTRACTION

Applying gas extraction to substances of unknown or large distribution coefficients (for example, aqueous solutions of oxygen compounds with $K > 10^3$) requires a dramatic increase in the ratio of the volumes of the gas and liquid phases. This is achieved by employing dynamic conditions. Such a system passes a sufficiently large volume (from a few to 100 liters) of air or an inert gas through several milliliters of the analyzed solution as a continuous flow of small bubbles. The equilibrium vapor is analyzed both before and after purging the sample.

The basic characteristics and conditions of conducting a continuous gas extraction of a substance from a nonvolatile or practically nonevaporating solvent were discussed in connection with the extraction for the measurement of the distribution coefficients (see Section 1.2).

The determination of volatile substances in solutions using this type of head-space analysis is accomplished by two methods[22]:

1. Static analysis of the equilibrium vapor above the solution, carried out both before and after a certain volume of gas is bubbled through the solution.
2. Direct analysis of the gas flow passed through an investigated solution.

In the first case, the analysis is accomplished in three stages (Fig. 1.6):
a. Static determination of the concentration of the analyzed impurity in an equilibrium gas phase C_G (corresponding to the concentration of the substance in solution C_L) according to the initial calibration C_G vs. A_G.

50 THEORY OF GAS-CHROMATOGRAPHIC HEAD-SPACE ANALYSIS

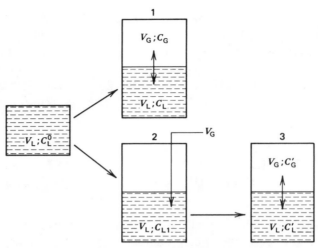

Figure 1.6. Schematic of the distribution of a substance between a liquid and a gas during the later stages of head-space analysis employing continuous gas extraction. See explanation in the text.

b. Bubbling pure gas of volume V_G through the initial solution of the substance having a concentration of C_L^0. With this action, the concentration of the substance decreases to C_{L1}.

c. Static determination of the concentration C_{L1} by the amount of substance in the equilibrium gas phase is completed by measuring the peak area A'_G, which is proportional to the concentration C'_G.

Quantitatively these processes can be expressed in the following manner. The initial concentration C_L^0 is related to C_G through Equation (1.20), and to C_{L1} through a similar relationship:

$$C_{L1} = C'_G \left(K + \frac{V_G}{V_L} \right)$$

The values C_L^0 and C_{L1} related through Equation (1.36) can be written

$$C_L^0 = C_{L1} \exp\left(\frac{v_G}{KV_L} \right)$$

The resulting ratio C_L^0/C_{L1} in this expression can be substituted by the ratio of the corresponding concentrations of the substance in a gas phase C_G/C'_G. Under sampling conditions, when equal volumes of equilibrium gas are introduced into the chromatograph the gas-phase ratio can be substituted

1.4 CONTINUOUS GAS EXTRACTION

using the proportionality of peak areas or heights in the chromatogram. Thus,

$$\ln\left(\frac{A_G}{A'_G}\right) = \frac{v_G}{KV_L} \quad (1.58)$$

Substituting the value of K from Equation (1.58) into Equation (1.20) results in the basic formula for the determination of the volatile components in a solution using a nonvolatile solvent by the HSA method of continuous gas extraction:

$$C_L^0 = C_G\left(\frac{v_G}{V_L \ln (A_G/A'_G)} + \frac{V_G}{V_L}\right) \quad (1.59)$$

In the second case, when analyzing the gas flow passed through an investigated solution, which is the same technique used for determining the distribution coefficient, it is necessary to measure the peak areas or heights A'_G and A_G in the chromatogram. These values are proportional to the equilibrium concentrations C_G and C'_G of the gas flow during the time that the gas volumes v_G and v'_G are bubbled through the solution. The value of A_G obtained from the preliminary calibration C_G vs. A_G allows calculation of an absolute value of C_G. Having these data, the initial concentration of a substance in a solution can be determined using the equation

$$C_L^0 = C_G \frac{v'_G - v_G}{V_L \ln (A_G/A'_G)} \exp\left[\frac{v'_G \ln (A_G/A'_G)}{v'_G - v_G}\right] \quad (1.59a)$$

which was obtained by substituting the values of C_L from (1.3) and K from (1.38) into (1.36).*

In employing gas extraction for the determination of K, the correct selection of the optimum value of the bubbled gas v'_G is of prime importance to the analysis. The dependence of $(\Delta C_L^0)/C_L^0$ on the conditions of the analysis (similar to that discussed in Section 1.2) leads to the desirability of retaining the condition $KV_L \leq v_G$. Under this condition an excessive increase in the volume of bubbled gas must be avoided, as it leads to a decrease in the area A'_G, increasing the error of its measurement and decreasing the detection limit. If the maximum value v'_G is experimentally

---

*Equation (1.59) is applicable only in cases when the volume of a gas space above a liquid in a container for extracting a substance from a solution is sufficiently small in comparison with the volume of the solution (for details, see Section 5.3).

limited, then the dependence of the error on the analytical conditions in the determination of C_G^0, assuming satisfactory accuracy of the analysis, allows the evaluation of limited values of K.

Continuous gas extraction of a substance from a solution in a volatile solvent

The process of continuous gas extraction and its analytical applications were discussed above for nonvolatile or practically nonvolatile solvents. In these cases evaporation of the basic solvent was excluded by preliminary saturation of a gas by its vapors. The principle of gas extraction applied to solutions of volatile substances in volatile solvents can considerably widen the possibility of practical applications.

Let us discuss the general case of changes in the concentration of the volatile substance in a solution of the volatile solvent when an inert gas is bubbled through it.[27] Consider a microbubble of an inert gas that has passed through a liquid and is saturated with its vapors. At the moment of its emergence from the liquid, the bubble has the volume dv_G. Due to evaporation, the solution volume V_L will change by dv_L, and the concentration of the substance in the liquid C_L will be $C_L + dC_L$. According to the distribution law (1.3) the equilibrium concentration of a substance in a gas bubble at its emergence from the liquid will be

$$C_G = \frac{C_L + dC_L}{K}$$

Then the equation of the mass balance in the substance for the fundamental passage of a microbubble of a gas can be written as

$$C_L V_L = (C_L + dC_L)(V_L + dV_L) + \frac{C_L + dC_L}{K} dv_G \qquad (1.60)$$

After the escape of the gas volume v_G (including the volume of evaporated liquid) from a liquid, initially assumed to be saturated, the volume of the solution V_L^0 will decrease due to evaporation by Fv_G. Here F is a measure of the volatility of a solvent having a vapor pressure P_L, a density d_L, and a molecular mass M at temperature T.

$$F = \frac{P_L M}{RT\, d_L}$$

1.4 CONTINUOUS GAS EXTRACTION

Thus,

$$V_L = V_L^0 - Fv_G \tag{1.61}$$

and

$$dV_L = -F\, dv_G \tag{1.61a}$$

These values of V_L and dV_L are substituted into the equation of mass balance (1.60). After the separation of variables and upon integrating over the volume of gas in the limits from 0 to v_G and over the concentration limits from C_L^0 to C_L, one obtains

$$C_L = C_L^0 \left(\frac{V_L^0 - Fv_G}{V_L^0}\right)^{(1-FK)/FK} \tag{1.62}$$

It must be noted that the derivation does not consider deviations from ideality of the gas or the solution. Effects related to the curvature of the surface of bubbles and hydrostatic pressure are also neglected. The solubility of the bubbled gas in the liquid and the limitations imposed by the diffusion processes are also disregarded. For low C_L (i.e., when analyzing the admixtures), the assumptions are justified, since the effect of these factors upon the relationship of C_G with C_L is relatively small and in most cases falls within the limits of the error in the measurements (on the order of several percent).

The derived equation describes the process of the elimination of a substance from its solution in a volatile solvent by purging with a pure gas. Further, it relates the immediate concentration in a solution to its initial volume, initial concentration C_L^0, the volatility of the solvent F, and the value of the distribution coefficient.

A similar problem was solved by Burnet[21] for solutions with a limited solvent volatility. The approximate equation obtained by this author in the designated system has the form

$$C_L = C_L^0 \left(\frac{V_L^0 - Fv_G}{V_L^0}\right)^{1/FK} \tag{1.63}$$

Equation (1.63) is applicable only under the conditions where $FK \ll 1$. Here the difference in values of C_L computed from Equations (1.62) and (1.63) must fall within the error of determination of the concentration of the substance in solution.

The ratio of the concentration of the purged solution to the initial concentration is

$$X = \frac{C_L}{C_L^0}$$

and the ratio of the volume of the dilute volatile solution to the initial volume after purging is

$$Y = \frac{V_L^0 - Fv_G}{V_L^0}$$

Equation (1.62) can be rewritten in the form

$$X = Y^{(1-FK)/FK} \qquad (1.64)$$

Investigation of Equation (1.62) illustrates the characteristics of gas extraction from volatile solvents as being quite considerable for the effective application of this process. The main parameter that determines the course of a gas extraction is the value of FK, the relationship between the volatilities of the dissolved substance and the solvent. The shape of the curves due to the change in the concentration of solutions with different values of FK is shown in Figure 1.7.

1. When $FK = 1$, the bubbling of a gas through a volatile solution does not change the composition; the solution behaves as an azeotrope. The straight line where C_L is a constant divides the C_L–v_G diagram into two areas, corresponding to solutions having linear and nonlinear relationships in the volatility of the components.

2. When $FK > 1$ (the upper field of the diagram), the process of gas extraction enriches the remaining solution with the dissolved substance; the gas extracts the volatile solvent to a relatively higher degree. Concentration C_L increases from its initial value C_L^0 along the concave curves.

3. Enrichment of the solution with the solvent can occur only when $FK < 1$ (the lower field of the diagram), i.e., when $K < 1/F$. This criterion allows the determination of the possibility of removing a particular substance by gas extraction from a given volatile solvent. For example, in aqueous solutions at 20°C ($F = 1.73 \times 10^{-5}$), K must be less than 57,800. In benzene solutions the maximum value of K is 2730.

4. If $FK = 0.5$, then Equation (1.62) describes the straight line that divides the convex ($0 < FK < 0.5$) and concave ($0.5 < FK < 1$) curves.

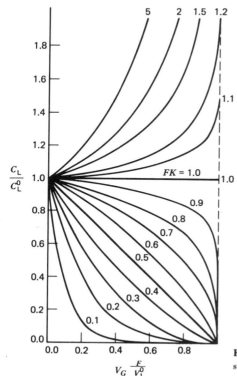

Figure 1.7. Concentration change of a substance in volatile solvents during the process of continuous gas extraction at various values of FK.

5. In the case of a nonvolatile solvent, $F = 0$. In this case, Equation (1.62) acquires the form (1.36), which was discussed earlier in connection with the utilization of continuous gas extraction for measuring distribution coefficients and the determination of volatile substances in nonvolatile solvents.

The important areas of analytical application of gas extraction using volatile solutions are:

1. Quantitative determinations by head-space analysis in the case of systems with known or unknown distribution coefficients
2. Extraction of volatile substances from solutions for their subsequent collection (for example, by adsorption or low-temperature cooling) for further investigation

3. Removal of solvent to enrich the remaining solution in the dissolved substance

Quantitative determination of volatile substances. Head-space analysis with continuous gas extraction for volatile solvents can be carried out by two methods. These have been discussed in the case of a practically nonvolatile solvent. The simplest case is the direct analysis of the gas. This method is applicable when the volume of the gas space over the liquid and the sample taken from it for analysis are sufficiently small compared to the volume of the solution V_L and, also, in comparison with v_G. Under these conditions the concentration of the component being determined in the gas phase C_G is related to the concentration of the equilibrium solution (in this case, the initial concentration) C_L^0 by the distribution law.

For the determination of substances with unknown values of K, the bubbling of a certain volume of gas v_G' through the solution is satisfactory. By measuring the peak areas or heights on the chromatogram, A_G and A_G', the distribution coefficient can be computed.

$$K = \frac{\ln Y}{F \ln (XY)} \tag{1.65}$$

(Here $X = A_G'/A_G$.) The initial concentration of the substance in solution is obtained from the formula

$$C_L^0 = C_G \frac{\ln Y}{F \ln (XY)} \tag{1.66}$$

For maximum accuracy the conditions of the analysis are dictated by the conditions for the measurement of K and can be obtained from similarly discussed examples (see Section 1.2).

Removal of volatile substances or solvent from a solution. The fraction of extracted substance remaining in solution is equal to the product of the volume fraction of the remaining solution Y and the fraction of initial concentration X. Therefore, the degree of extraction Z is

$$Z = 1 - XY$$

Considering (1.64),

$$Z = 1 - Y^{1/FK}$$

The volume of gas necessary to purge the given volume of solution to obtain the assigned degree of extraction of the volatile substance will be equal:

$$v_G = \frac{V_L^0}{F} [1 - (1 - Z)^{FK}] \qquad (1.67)$$

For example, it is necessary to bubble 30 ml of air through 1 ml of aqueous solution at 20°C ($K = 6.5$, $F = 1.73 \times 10^{-5}$) to almost completely ($Z = 0.99$) extract benzene. The removal of traces of ethanol ($K \simeq 5000$) from 1 ml of aqueous solution requires 10 liters of gas, while one-third of the water will evaporate.

The basic equation (1.62) allows the evaluation of the degree of enrichment of a solution by the substance and the selection of optimum conditions for conducting the gas extraction—that is, the initial volume of solution V_L^0 and the volume of purged gas necessary for achieving the assigned value of X.

It must be noted that the foregoing formulas for various applications of continuous gas extraction retain their form for nonequilibrium extraction processes. However, the values of K in such cases will differ from the distribution coefficients and will include the effects of the rates of mass transfer.

1.5 METHODS OF INCREASING THE SENSITIVITY OF HEAD-SPACE ANALYSIS

The determination of trace volatile substances in solutions is the most popular application of head-space analysis. Therefore, the sensitivity of the method quite frequently plays an important role in the evaluation of its possible application.

In practice, sensitivity of the analysis is increased in two basic ways: (1) by lowering the values of the distribution coefficients of the analyzed substances in the investigated system and (2) by adsorbing or cryogenically trapping the substances contained in the equilibrium gas before introducing them into the chromatographic column. The method of increasing sensitivity of the gas-chromatographic portion of the analysis at the expense of using highly efficient columns, selective detection, and so on, are well known.

58 THEORY OF GAS-CHROMATOGRAPHIC HEAD-SPACE ANALYSIS

These are described in sufficient detail in the specialized literature and are not discussed here.

The value of the distribution coefficients can be decreased by increasing the temperature of the system during the process of establishing equilibrium, lowering the solubility of certain substances, or converting them into more volatile and less soluble compounds.

The increase in the degree of sensitivity achieved by decreasing the value of the distribution coefficient from K to K' under the conditions of a constant phase-volume ratio, is determined by the relationship

$$\alpha = \frac{C'_G}{C_G} = \frac{K + V_G/V_L}{K' + V_G/V_L} \qquad (1.68)$$

This equation indicates that such a method is preferred only when $K \gg V_G/V_L$.

The effectiveness of the thermal method of increasing the sensitivity of head-space analysis is determined by the temperature dependence of the distribution coefficient (discussed in Section 1.1) and usually does not exceed the order of 1. The limiting factor in this case is the experimental complexity of conducting the analysis at temperatures above 80–100°C.

For HSA determination of organic substances in sufficiently volatile solvents (alcohol, acetic acid, dimethyl formamide), a considerable increase in the temperature during the establishment of phase equilibrium is undesirable. Such an increase would lead to a sharp increase in the concentration of the solvent in the gas phase, which would result in a large peak in the chromatogram overlapping the peaks of the analyzed substances. In these cases the best results are obtained by diluting the solution with water. Thus, according to the data in Table 1.4,[8] dilution of an acetate solution of aromatic hydrocarbons with water sharply decreases the values of the distribution coefficients. Such a method can lead to a considerable increase in the sensitivity of the analysis only for the substances that are less soluble in water (higher alcohols, ethers, nitriles, hydrocarbons, etc.). The sensitivity of the analysis increases as the solubility of the substance in water decreases. An example illustrating such regularity is the measurement of the concentration in the gas phase of normal alcohols (C_2–C_5) above their solutions in aqueous dimethyl formamide (Fig. 1.8).[28] The data indicate that the degree of increase in the sensitivity of head-space analysis (expressed as the slope of the straight lines) increases in a homologous series when diluting the solvent

1.5 INCREASING THE SENSITIVITY OF HEAD-SPACE ANALYSIS

Table 1.4 Distribution Coefficients of Benzene, Toluene, and m-Xylene in Aqueous Acetic Acid Solutions at 25°C[a]

Concentration of Acetic Acid, %	Distribution Coefficient		
	Benzene	Toluene	m-Xylene
100	773	2100	5500
95	590	1600	3830
90	450	1160	2670
85	350	840	1860
80	270	610	1300
75	205	440	900
65	123	224	458
55	71	118	225
45	42	62	111
30	18.9	23.8	38.2
25	14.5	17.3	26.8
20	11.1	12.5	18.8
15	8.5	9.1	13.2
10	6.6	6.6	9.2
5	5.0	4.8	6.5
1	4.1	3.7	4.9

[a]Computed by the interpolation of experimental values of the distribution coefficients.[8,13]

with water. This is the result of a greater decrease in the value of K' in comparison with K for succeeding members of the homologous series of alcohols.

When selecting the degree of dilution, the increase in the volume of the solution relative to the decrease in the sensitivity of the analysis must be considered. If, as a result of diluting a solution, the distribution coefficient decreases from K to K', the volume of solution increases from V_L to V'_L, and the value V_G remains constant, then the degree of concentration of the substance in the gas phase is determined by the relationship

$$\alpha = \frac{K + V_G/V_L}{K + V_G/V'_L} \frac{V_L}{V'_L} \quad (1.69)$$

This allows the evaluation of the optimum degree of dilution, when the dependence of the distribution coefficient on the relative amounts of the

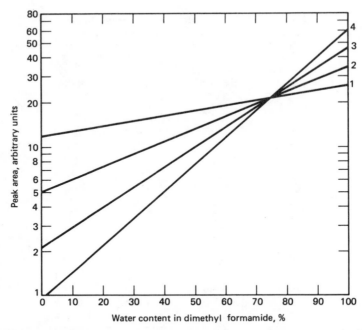

Figure 1.8. Relationship of the chromatographic peak areas to the water content of dimethyl formamide solutions. The peak areas are measured from the equilibrium vapor above the solutions of normal C_2–C_5 alcohols (120 mg/liter each). *1*, Pentanol; *2*, butanol; *3*, propanol; *4*, ethanol.

organic solvent and water in the liquid phase is known.* Figure 1.9 illustrates the dependence of the enrichment of aromatic hydrocarbons in the gas phase on the amount of water in aqueous acetic acid solutions of these substances, as computed from the data in Table 1.4. It is evident from the graph that the maximum degree of enrichment (at 12, 29, and 56 times for benzene, toluene, and *m*-xylene respectively) is reached when glacial acetic acid is diluted 5–6 times. Further dilution of the solution leads to a sharp decrease in the concentration of the substance in the equilibrium gas, since the decrease in the distribution coefficient is not compensated by the increase in the volume of the solution.

In connection with the greatest effect obtained in the dilution of organic solvents with water for less soluble substances, one must consider the possibility of forming emulsions. The latter can lead to gross errors in the

---

*This dependence can be determined from Equation (1.19) when the values of the distribution coefficients are known for the pure solvent and water.

1.5 INCREASING THE SENSITIVITY OF HEAD-SPACE ANALYSIS 61

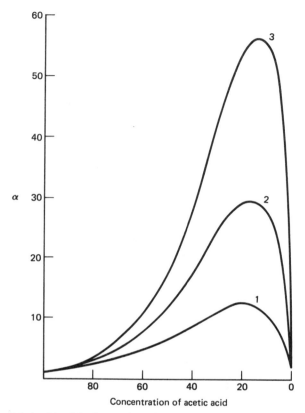

Figure 1.9. Relationship of the degree of enrichment of the gas phase to the water content of acetic acid solutions. The initial value of $V_G/V_L = 10$. *1*, Benzene; *2*, toluene; *3*, *m*-xylene.

analysis, since the partial pressure of the vapor of the analyzed component over the solution of the emulsion will be equal to the vapor pressure of the pure substance.

Besides decreasing the value of K, another advantage must be attributed to water dilution, the decrease in the vapor pressure of the primary solvent. This fact considerably facilitates, and in certain cases accelerates, the performance of the chromatographic analysis at the expense of a considerable decrease or even the elimination of the solvent peak.

By introducing mineral salts into the solution, one can achieve a decrease in the distribution coefficient and also an increase in the sensitivity of headspace analysis during the analysis of aqueous solutions of volatile organic substances. With organic solvents, the gain in sensitivity is obtained at the expense of a decrease in the solubility of the substance.

The nature of the salt used influences the degree of increase in the sensitivity of head-space analysis. The results of the relative measurements of the increase of the concentration of ethanol in the vapor phase over a 2% ethanol–water solution before and after the introduction of equal quantities of various salts are given in Table 1.5.[29] The concentration of a salt can also have an effect since maximum saturation of the solution is not always justified. Usually a sufficiently sharp decrease in the distribution coefficient is observed with salt concentrations up to 10–20%. Further increases in the salt content of the solution decrease the value of K only slightly and lead to an excessive waste of the reagent.

The use of mineralization of the investigated aqueous solution to increase the sensitivity of head-space analysis has obtained wide application in the analytical practice. It is especially popular for determining volatile organic substances in biological materials, food products, and sewage. The greatest benefit is being achieved in the analysis of polar substances. According to Nelson and Hoff,[13] the addition of sodium sulfate to water at 28°C in the ratio of 0.6:1 decreases the distribution coefficients of acetone, ethyl acetate, and methyl propyl ketone by 11, 16, and 19 times, respectively. Introduction of magnesium sulfate into an acetic solution results in the reduction of the minimum detectable limit of aliphatic fatty acids by more than an order of magnitude.[30].

Table 1.5 Increase in the Concentration of Ethanol in the Gas Phase upon Addition of Various Salts to the Aqueous Solution at 60°C

Salt	Degree of Increase of Concentration in the Gas over the Solution[a]
Ammonium sulfate	5
Sodium chloride	3
Potassium carbonate	8
Ammonium chloride	2
Sodium citrate	5

[a]Numbers represent the increase relative to the concentration of ethanol in the gas phase over a 2% ethanol-water solution without the addition of any salt.

1.5 INCREASING THE SENSITIVITY OF HEAD-SPACE ANALYSIS

A promising method of increasing the sensitivity of head-space analysis is the conversion of the substances to be determined into more volatile and less soluble derivatives. This method is applicable to the determination of reactive compounds, characterized by large ($> 10^3$) distribution coefficients. These include organic acids and alcohols in aqueous solution. By direct head-space analysis the detection limit of these substances is at concentrations of 10^{-1}–10^{-4}%. However, by converting the organic acids into methyl esters by their reaction with dimethyl sulfate, the minimum detectable limit can be improved: for example, the limit of detection of trichloroacetic acid is reduced to 10^{-5}%.[31] For alcohols, either the haloform reaction (for trichloroethanol[31]) or their conversion into the ethers of nitrous acid are used.[32,33] In this case, the limit of detection becomes 10^{-5}%.

The sensitivity of head-space analysis can be considerably increased by increasing the volume of the equilibrium gas introduced into the chromatographic column. The efficiency of the column will not be diminished if the substances contained in the equilibrium gas are retained on an adsorbent or are passed through a cryogenic trap before being fed into the chromatograph. The degree of increase in the sensitivity when an intervening concentrator is used depends upon the volume of the equilibrium gas conducted through it. Realistically, the increase can reach two to four orders of magnitude.

Head-space analysis combined with preliminary concentration is successful in two cases: when the gas extraction of substances from a solution is done under static conditions and when it is done under dynamic conditions using a flow of pure gas. The latter case is more effective since a better degree of enrichment is achieved although great volumes of extracting gas are consumed. Also, in continuous extraction, a greater quantity of the substance is extracted with the same volume of gas. Only single cases are known of the use of gas extraction under static conditions employing adsorption[34] and cryogenic[35] concentration of the analyzed substances. However, the combination of evaporating a substance from a solution accompanied by its concentration before it is fed into the chromatograph column has gained wide acceptance (see Section 3.1). Nevertheless, to completely exclude the possibility of concentrating the substances from an equilibrium gas, obtained under static conditions, is not justified. Technically this method is simpler for substances with small K (< 10) and allows the determination of concentrations as low as 10^{-8}–10^{-9}% when extracting the substance from 10–100 ml of the equilibrium gas.

Combining continuous gas extraction with the concentration of the substance before its admittance to the chromatographic column allows the same detection limit for substances with large K. For this reason it is necessary to use considerably large volumes of the equilibrium gas. In this variation of head-space analysis two considerations are necessary when selecting the conditions and completing the analysis.

1. The degree of increase in analytical sensitivity depends upon the quantity of the substance extracted from the solution (Z). This quantity is determined by the conditions under which the continuous gas extraction is conducted: the quantity of initial solution, its temperature, and the volume of purged gas. Optimum conditions can be selected by consulting the relationships given in Section 1.4. For example, in the case where $K = 10^3$, if a detection limit in the order of $10^{-8}\%$ is to be achieved and the volume of the aqueous solution is 10 ml, then, according to Equation (1.36) 29 liters of gas must be bubbled through the solution in order to almost completely extracting ($Z > 0.95$) the substance from the solution. However, when the admixture is extracted from 100 ml of solution, the same detection limit can be achieved by bubbling only 10 liters of pure gas through the solution.

2. The specifications in the selection of an absorber are quite important. In the case of cryogenic accumulation, these specifications refer to the construction of the trap (nature and volume of the adsorbent) and the temperatures required for the complete adsorption of the substances being analyzed and their subsequent quantitative desorption. The conditions under which the substance eluted from solution with the gas flow is accumulated in the trap depend on the properties (boiling point and polarity) of both the substance being analyzed and the solvent. These properties are considered in Section 3.1.

As a conclusion we may state that the intervening accumulation of a substance when using head-space analysis techniques allows a considerable increase in the sensitivity of the analysis. However, this variation has inherent shortcomings in the adsorption and cryogenic collection and concentration methods. Therefore, the use of this method for the analysis of easily oxidized and thermally unstable substances must be carefully considered.

REFERENCES

1. A. B. Littlewood, *Gas Chromatography*, 2nd ed., Academic Press, New York, 1970, p. 46.
2. E. B. Sandell and T. S. West, *Pure Appl. Chem.*, **51**, 43 (1979).
3. G. T. Preston and J. M. Prausnitz, *Ind. Eng. Chem. Fundamn.*, **10**, 389 (1971).
4. V. B. Kogan, *Heterogeneous Equilibria*, L. Khimiya, 1968, 432 pp.
5. V. N. Borisov, N. G. Zenkevich, and N. A. Smirnova, *Vestn. Leningr. Univ.*, **10**, 135 (1974).
6. L. Rohrschneider, in *Advances in Chromatography 1973* (A. Zlatkis, Ed.), p. 189.
7. V. V. Tsibulskii, I. A. Tsibulskaya, and N. N. Yaglitskaya, *Zh. Analit. Khim.*, **34**, (1979).
8. I. A. Tsibulskaya, doctoral dissertation, Leningrad University, 1979.
9. A. G. Vitenberg and M. I. Kostkina, *Zh. Analit. Khim.*, **34**, 1800 (1979).
10. A. G. Vitenberg and M. I. Kostkina, *Vestn. Leningr. Univ.*, **4**, 10, (1980).
11. S. A. Smirnova, Novye V., Kand. Diss. Leningrad University, 1978, 124 pp.
12. V. V. Tsibulskii, A. G. Vitenberg, and I. A. Khripun, *Zh. Analit. Khim.*, **33**, 1184 (1978).
13. P. E. Nelson and J. E. Hoff, *J. Food Sci.*, **33**, 479 (1968).
14. I. R. Krichevskii, *Zh. Fiz. Khim.*, **9**, 41 (1937).
15. E. W. Tiepel and K. E. Gubbins, *Can. J. Chem. Eng.*, **50**, 361 (1972).
16. C. McAuliffe, *Chem. Technol.*, **1**, 46 (1971).
17. A. G. Vitenberg, B. V. Ioffe, Z. St. Dimitrova, and I. L. Butaeva, *J. Chromatogr.*, **112**, 319 (1975).
18. C. Feldman, *Anal. Chim. Acta*, **96**, 383 (1978).
19. R. A. Hasty, *Can. J. Chem.*, **46**, 1643 (1968).
20. I. H. Williams and F. E. Murrey, *Pulp Paper Mag. Can.*, **1966**, 347.
21. M. G. Burnet, *Anal. Chem.*, **35**, 1567 (1963).
22. A. G. Vitenberg and B. V. Ioffe, *Dokl. Akad. Nauk SSSR*, **235**, 1071 (1977).
23. B. V. Ioffe and A. G. Vitenberg, *Chromatographia*, **11**, 282 (1978).
24. A. G. Vitenberg, B. V. Stolyarov, and S. A. Smirnova, *Vestn. Leningr. Univ.*, **16**, 132 (1977).
25. J. Novák, *Quantitative Analysis by Gas Chromatography*, Marcel Dekker, New York, 1975, p. 107.
26. J. Drozd and J. Novák, *J. Chromatogr.*, **152**, 55 (1978).
27. A. G. Vitenberg and B. V. Ioffe, *Dokl. Akad. Nauk SSSR*, **288**, 352 (1978).
28. H. Hachenberg and A. P. Schmidt, *Gas Chromatographic Head-space Analysis*, Heyden, London, 1977, p. 12.

29. G. Machata, *Clin. Chem. Newsletter*, **4**(2), 29 (1972).
30. B. Kolb, *J. Chromatogr.*, **122**, 553 (1976).
31. G. Triebig, Vorträge zum 2. Internationalen Colloqium über die Gas-chromatographische Dampfraumanalyse in Überlingen, October 18-20, 1978. Bodenseewerk Perkin-Elmer, GmbH.
32. P. K. Gessner, *Anal. Biochem.*, **38**, 499 (1970).
33. B. Komers and Z. Sir, *J. Chromatogr.*, **119**, 251 (1976).
34. M. Gottauf, *Z. Anal. Chem.*, **218**, 175 (1966).
35. R. E. Hurst, *Analyst*, **99**, 302 (1974).

CHAPTER TWO
Instrumentation for Head-Space Analysis

2.1 BASIC METHODS OF INTRODUCING THE VAPOR PHASE INTO THE CHROMATOGRAPH

The operation that determines the accuracy of the quantitative analysis is the introduction of the gas that is in equilibrium with the condensed phase (liquid or solid) into the chromatograph column. This process differs greatly from the common methods of introducing gas samples. It requires the use of special techniques and procedures that are dictated by the properties of the heterogeneous gas–condensed phase system.

The compulsory condition of sample preparation for the analysis is the maintenance of a constant and well-reproducible temperature in an equilibration container. Besides that, the process of sample introduction must ensure that there is no loss of the substance in the gas phase of the container. Therefore, the instrumentation must secure the shortest time of sample preparation for the analysis and the method selected for introducing the gas should ensure that the equilibrium in the liquid–gas system is maintained and there is no loss in the analyzed substances. Special attention must be given to the possibility of the saturated vapors condensing on the connecting parts in the gas flow system, which are at temperatures lower than that of the solution investigated. Changes in pressure can lead to a loss of equilibrium in the system and correspondingly change the concentration of the analyzed components in the gas phase.

The devices used in the static variations of head-space analysis, for introducing the equilibrium gas into a chromatograph can be divided into two basic groups. One type uses constant-volume containers for the establishment of equilibrium. Samples taken from these containers are generated at varying pressures. The other type of device is applied to systems of variable gas-phase volume, and sampling is carried out at constant pressure. The principle of constant pressure is realized in practical applications in all the dynamic variations of head-space analysis.

Each of these types of device has certain advantages and limitations that should be considered when selecting the method of sample preparation, type of quantitative analysis, and method of introducing the gas sample into the chromatograph. These considerations depend upon actual analytical conditions (volatility of main solvent, magnitude and constancy of the distribution coefficient in different samples, required sensitivity, accuracy of the analysis, etc.).

2.1 BASIC METHODS OF INTRODUCING THE VAPOR PHASE

Numerous implements for introducing the equilibrium gas sample into a chromatograph are described by Hachenberg and Schmidt[1]. Therefore, only typical variations and designs with their classification and characteristics will be discussed here. New methods are presented in more detail, as well as commercially available instrumentation.

In static systems the direct introduction of the equilibrium gas into the chromatograph from a constant-volume container can be achieved by several methods:

1. Sampling with a gas syringe followed by introduction of the sample into the vaporization chamber of the chromatograph. The syringe temperature must be equal to or higher than that of the equilibration container. If it is not, the vapors of the solvent may condense, and the concentration of the analyzed components in the gas will change sharply. In order that the gas in the container and in the syringe remain at atmospheric pressure,* after obtaining a sample remain the syringe (before introducing the needle into the container) is filled with air (or inert gas), the volume of which is equal to that taken from the container. Introducing a volume of pure gas into the vessel leads to a decrease in the equilibrium concentration of the substance in the gas phase. Therefore, the sampling must be done with multiple injections by the syringe over a time sufficient to reestablish equilibrium in the system.

The main advantage of the method is its simplicity and the possibility of using standard gas chromatographic instrumentation. Among the considerable disadvantages is the loss of the substances being determined by adsorption onto the walls of the syringe and especially onto the surface of the septum. The latter must be considered when analyzing substances easily absorbed by rubber (for example, hydrocarbons or amines). In addition, when a syringe is used to introduce the gas, it is difficult to obtain good reproducibility of peak heights on the chromatogram because of backpressure in the gas flow system of the chromatograph, irregular introduction of a sample into a column, and possible leaking of the seal of the vaporization chamber.

2. The introduction of a sample into a chromatograph column through a gas sampling valve the metered volume of which is filled by changing the

---

*If this condition is not fulfilled, when the needle is removed from the container the sample in the syringe will be diluted by air, and the pressure of equilibrium gas over the liquid will decrease.

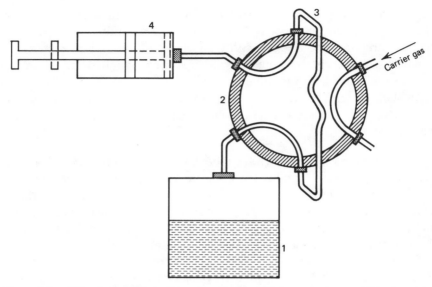

Figure 2.1. Filling the sample loop of the valve with an equilibrium gas, using a syringe as the pump. *1*, Container for establishing equilibrium; *2*, gas sampling valve; *3*, sample loop of the valve; *4*, syringe.

pressure in the system. Figure 2.1 illustrates a case in which the sample is introduced by means of repeated short decreases of the pressure in the system. A hypodermic syringe is used as a pump and is connected to the exit port of the valve.* Each injection by the syringe fills the loop of the valve with gas from the container. The use of a gas sampling valve which is thermostated at high temperatures (up to 100–200°C) considerably lessens (but does not eliminate) adsorption losses of the substance. This allows the attainment of high reproducibility in the introduction of gas samples into the column (at the level of tenths of a percent). However, this method does not exclude the possibility of a reduction of the concentration of the substance in the gas phase in the container during filling of the sample loop of the valve. In order to lessen this effect, the magnitude of the excessive pressure that develops during pumping with the syringe should not exceed

---

*This method has been applied to the analysis of waste water from the manufacture of polyvinyl chloride (N. R. Litvinov, T. M. Lyutova, Yu. G. Sukhareva, Ochistka Promyshlennykh vybrosov i tekhnika bezopasnosti na khimicheskikh predpriyatiyakh, NIITEKhIM, Moscow, Ref. sb, No. 9, 1976, p. 17).

2.1 BASIC METHODS OF INTRODUCING THE VAPOR PHASE   71

variations of this method, based on the development of negative pressure in the system (see, for instance, ref. 2).

3. Pneumatic introduction of the equilibrium gas[3,4] is accomplished by admitting an inert gas into the equilibration container and developing a pressure greater than that at the inlet of the chromatographic column (Fig. 2.2). Connecting the gas space of the container with the vaporization chamber of the chromatograph (or with another point of the gas-flow system before the chromatographic column) secures quick introduction of the sample. Its volume is regulated primarily by the difference of the pressures the permitted error of the analysis (in fractions). There are other possible

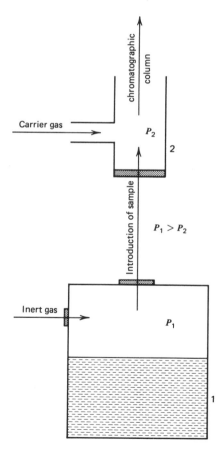

Figure 2.2. Pneumatic injection of the equilibrium gas into a gas chromatograph. *1*, Container for establishing equilibrium; *2*, injection port of the gas chromatograph.

in the container and in the gas-flow system of the chromatograph. In automated systems developed by Perkin-Elmer[5] (the first company to apply this principle), the pneumatic method of sampling is realized in a different manner. The container with the sample is connected to the gas-flow system of the chromatograph at the inlet of the column. After equalizing the pressure, the supply of the carrier gas is briefly interrupted. This causes a decrease in the pressure in the gas line and develops a pressure differential between the sample container (P_2) and the inlet of the chromatographic column (P_1). In this case the volume of a sample is regulated by the time the carrier gas supply is interrupted. (Pneumatic sampling systems are discussed in detail in the following sections which describe the special attachments and automatic head-space analyzers.)

The basic advantage of this method lies in the pulsed introduction of a sample directly from the gas space of a container into the chromatographic column. The elimination of intermediate assemblies completely excludes adsorption losses and the "memory" of the sampler. If the excessive pressure in the container is constant, high reproducibility of the sample volume is achieved but its absolute volume remains unknown.

Selecting samples from containers with constant gas space volume is characterized by changing the initially established concentration of the substance in a solution and the gas above it. Therefore, the possibility of repeated introduction of equilibrium gas is limited by the parameters of the analyzed system (by the value of the distribution coefficient and the relationship of the phase volumes) and by the quantity of the selected sample. Quantitatively, these limitations can be expressed in the following manner.[6,7]

If a sample of volume v_g is collected from the gas phase of an equilibrium system with fixed volumes of liquid V_L and gas V_G, and the drop in the pressure of the system is compensated by introducing an equal volume of a pure gas, then the concentration of a substance in the gas phase C_G^1 after removal of a single sample in comparison with the initial concentration C_G, assuming the distribution coefficient is constant, will be

$$C_G^1 = C_G \left(1 - \frac{v_g}{KV_L + V_G}\right) \qquad (2.1)$$

For n samplings,

$$C_G^n = C_G \left(1 - \frac{v_g}{KV_L + V_G}\right)^n \qquad (2.2)$$

2.1 BASIC METHODS OF INTRODUCING THE VAPOR PHASE 73

The change in concentration upon repeated sampling of the gas in equilibrium with a solution can be disregarded if the relative change in concentration does not exceed the required reproducibility in the detector's response (the peak areas or heights in the chromatograms) upon the consecutive introduction of a gas into the chromatograph with a constant concentration of a substance, that is, if

$$\frac{\Delta C_G}{C_G} = \cdots = \frac{\Delta C_G^n}{C_G^n} \geq \frac{C_G - C_G^n}{C_G} \tag{2.3}$$

From Equations (2.2) and (2.3) it follows that the limitations of the repeated selection of samples from a container can be extended to substances not only with small, but also with medium values of K. For example, when $V_G = V_L$, $v_g = 0.1 V_L$, and the error of determination is $C_G < 1\%$, the second sampling of the system is permissible if $K \geq 9$. At $V_G/V_L = 10$, with other conditions remaining constant, one can analyze substances that have a distribution coefficient of 90 or greater. If the minimum number of samples is 10, then the minimum value of the distribution coefficient increases to 490.

These examples indicate that when volatile and poorly soluble substances are determined, one must use systems with constant volume and take into consideration the given relationships. Besides this, when analyzing liquid samples, the application of systems with a constant gas-space volume considerably increases the difficulty in applying the variations of the HSA method. These variations require the substitution of an equilibrium gas by a pure gas such that in the process of analysis the numerical values of K characteristic for a given sample are determined.

Static systems with variable gas-space volume do not have these limitations, and it is possible to collect samples without changing the concentration of a substance in the contacting phases. Therefore, the number of repeated measurements of the equilibrium concentration of a substance in the gas phase is limited only by its volume.

Various methods can be employed to compensate for the decrease in the pressure in a system that occurs when samples are taken (or reduction in the volume of a gas phase):

1. A liquid can be added to constant-volume containers. In this case, the systems used lend themselves well to gas analysis (Fig. 2.3). The contents of a closed container are brought to atmospheric pressure by the use of a liquid

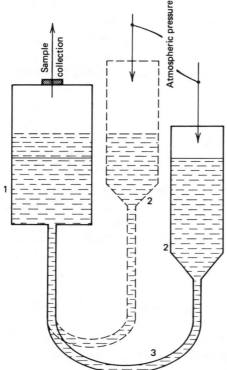

Figure 2.3. Equilibrium gas sampling from a system with variable-volume gas space at constant pressure. *1* Container for establishing equilibrium; *2* auxiliary container with the sample solution; *3* flexible tube.

flowing from an auxiliary container that is open to the atmosphere. In this system, atmospheric pressure is attained through the equalization of the liquid levels. The applicability of samples to head-space analysis lies in the ability of the analyzed substance to retain its concentration when the liquid is placed in the container to establish the equalization.

2. Equilibrium can be established in variable-volume containers. In this case, the initial pressure is maintained by taking samples from a container whose total volume decreases with sampling. Containers that can be used in this manner include plastic bags or small bottles, but are most frequently glass syringes.

The basic advantage of using variable-volume containers is the possibility of sampling the gas phase under conditions of thermodynamic equilibrium, that is, without any change in the concentration of a substance in one of the phases. Another advantage is the possibility of eliminating the errors related to the adsorption of a substance on the walls of the container, by considering certain corrections to the calculations (see Section 1.2). In addition, containers with variable volume allow total substitution of pure gas for the equilibrium gas. When necessary, both the analysis and the determination of the distribution coefficients can be carried out simultaneously. The shortcomings of variable-volume systems as compared to the pneumatic method are related to the possibility of sorption losses or to the appearance of a "memory" of the samples in the analysis of trace impurities.

In dynamic systems (continuous gas extraction), vapors of the analyzed substance are usually introduced directly into the chromatographic column with a thermostated gas sampling valve at 100–150°C. The sample loop is continually flushed by a flow of gas that has first been bubbled through the solution to be investigated.

Accumulation of a substance from an equilibrium gas by means of adsorption or cryogenic methods requires that the concentrate obtained be introduced quickly into the chromatograph by rapidly heating the entrainment separator. This is connected to the carrier-gas flow which enters the separation column. Typical versions of such arrangements are discussed in the following section.

2.2 LABORATORY EQUIPMENT FOR GAS CHROMATOGRAPHS

Due to their simplicity, availability, and relatively low cost, vessels normally used as standard laboratory glassware are widely used as constant-volume containers for the determination of the equilibrium distribution of a substance in a static state. Most frequently used are 5–40-ml glass vials sealed with a rubber septum. Sampling of a gas at equilibrium in a container and its injection into the loop of the gas sampling valve of a chromatograph are carried out with a gas-tight syringe.

The methods of injecting an equilibrium gas into standard chromatographs based on pneumatic principles were developed by Pauschmann[3] and Göke.[4] These methods require minor changes in the design of the chromatographs.

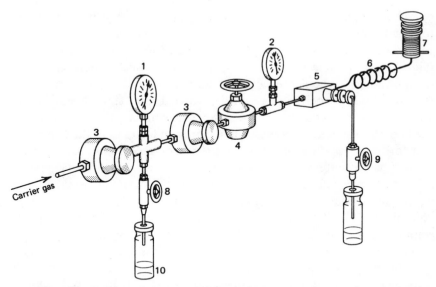

Figure 2.4. Schematic of the system for the injection of equilibrium vapor into the gas chromatograph, according to Pauschmann[3]. *1*, Manometer to measure the pressure in the container for establishing equilibrium (P_1); *2*, manometer to measure the inlet pressure at the chromatographic column (P_2); *3*, pressure regulators; *4*, flow controller; *5*, evaporation chamber of the gas chromatograph for liquid samples; *6*, chromatographic column; *7*, detector; *8*, *9*, valves; *10*, container with the sample.

The Pauschmann method requires the addition of a gas switching system (Fig. 2.4)* with two valves which end with needles that may be inserted through a rubber septum into a container to establish an equilibrium. Valve *8* is placed into a position where the pressure will exceed the pressure at the inlet to the chromatographic column, and valve *9* is connected to the vaporization chamber of the chromatograph. Sampling is done in the following manner: The needle in valve *8* is inserted through the rubber septum of the sample container; valve *8* is opened, and the carrier gas enters the sample container, establishing a pressure higher than the pressure at the vaporization chamber. Then valve *8* is closed, and the sample container is removed. After equilibrium is established at the higher pressure, the needle of valve *9* is inserted into the gas space above the sample solution in the

---

*The schematic has been somewhat altered from the original. In order to improve control of the pressure differential and increase the accuracy of the measurements, a manometer and a pressure regulator have been added.

2.2 LABORATORY EQUIPMENT FOR GAS CHROMATOGRAPHS   77

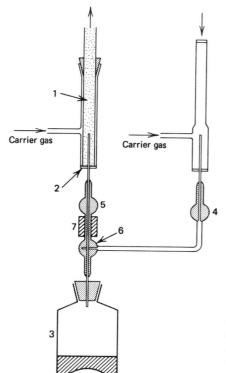

Figure 2.5. System to inject the equilibrium vapor into the gas chromatograph, according to Göke[4]. *1*, Chromatographic column; *2*, rubber septum; *3*, container with the sample; *4*, *5*, (metal) valves; *6*, three-way (metal) valve; *7*, evaporation chamber for liquid samples.

same container. When valve *9* is opened, the sample of equilibrium gas quickly flows into the vaporization chamber of the chromatograph.

The Göke method utilizes a dual-column gas chromatograph, in which only one column is used (Fig. 2.5). Stainless steel valve *4* allows the pressure in the sample container to become higher than the pressure at the inlet of the chromatographic column. To inject the sample, valve *4* is closed, the three-way valve *6* is turned 90°, and shutoff valve *5* is opened.

A simple apparatus was developed by Piorr[8] (Fig. 2.6) for a typical system with a variable-volume gas space. A pear-shaped glass bulb is hermetically sealed in the neck with a rubber septum and a screw-on cap. To level off the pressure, the outlet at the bottom of the flask is connected with flexible tubing to a glass valve. The entire system (including the flexible tube) is filled with the sample solution. The container is then hermetically sealed, and a required volume of inert gas or air is introduced into the system. After

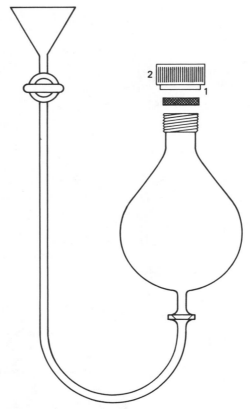

Figure 2.6. System with variable-volume gas phase, for sampling the equilibrium gas at constant pressure. *1*, Rubber septum; *2*, screw-cup.

adjusting the pressure in the system to atmospheric pressure and establishing an equilibrium, gas samples are taken with a gas-tight syringe and injected into the chromatograph. The samples can also be injected by means of a gas sampling valve. In such a case the equilibrium gas will be pushed out of the bulb by the sample solution and will be contained in the sample loop of the gas valve.

Another version of a system with a variable gas-phase volume, one that completely excludes the evaporation of the volatile components of a sample in the injector port, was recommended by Ronkainen[9] (Fig. 2.7). The fluid volume in a container is increased as the gas-sample equilibrium is established by fluid flowing in from another container in which the pressure is

2.2 LABORATORY EQUIPMENT FOR GAS CHROMATOGRAPHS 79

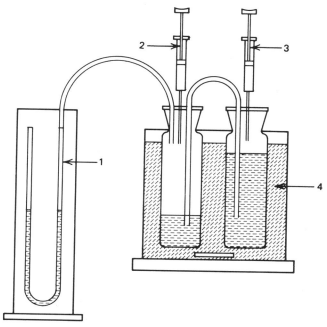

Figure 2.7. System for sampling with equalization of the pressure. *1*, Water manometer; *2*, syringe to equalize the pressure; *3*, syringe for sampling; *4*, thermostat.

equalized by injection of a gas from an auxiliary syringe. The pressure is measured by means of a water manometer.

For sampling an equilibrium gas at constant pressure, an apparatus developed to employ 20-, 50-, or 100-ml glass syringes (Fig. 2.8) is used successfully.[6,7] The syringe is sealed by a rubber septum that is fastened to the body *4* of the sealing device by a screw-on cap *7*. The body *4*, which is made of teflon, is fixed on the cannula *3* of a syringe by epoxy resin. The possibility of the sample leaking through a space between the plunger *1* and cylinder *2* of the syringe is avoided by a rubber washer *10* fastened by a union nut *11*. To maintain constant temperature, the syringe is mounted in a transparent plexiglas cover, which is connected to a circulatory thermostat. The plexiglas cover mounting is sealed by soft rubber washers *9*, *10*. Exact recording of volumes of fluid and gas introduced into the syringe is accomplished by the use of guides that limit the motion of the plunger *1*. The fluid or gas is introduced into the device from a standard medical syringe through

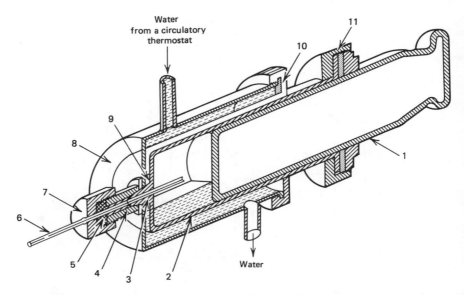

Figure 2.8. Variable-volume sampling device to inject the equilibrium gas at constant pressure into the gas chromatograph. *1*, Plunger of the gas-tight syringe; *2*, syringe cylinder; *3*, syringe cannula; *4*, body of the sealing device; *5*, rubber septum; *6*, capillary tube; *7*, screw-cup; *8*, Plexiglas cover; *9*, *10*, rubber washers; *11*, union nut.

the rubber septum *5*. During a process in which an equilibrium system in the apparatus must be maintained for a prolonged period, it is best to remove the capillary tube *6* to eliminate the loss of substance due to diffusion, and to replace it immediately before the analysis.

Gas sampling into a chromatographic column is accomplished by a gas sampling valve according to the schematic shown in Fig. 2.9. The inside space of the temperature-controlled syringe *1* is connected to a gas valve *2* by means of a stainless steel and teflon capillary tube (i.d. 0.5–0.8 mm). The capillary tube is inserted into the inside gas space through a rubber septum. The exit of the gas sampling valve *2* is connected directly to the chromatographic column *3* bypassing the vaporization chamber. The sample loop of valve *2* is filled with a gas sample by displacing the plunger of syringe *1*. The volume of gas passed through the sample loop is measured by a soap-film meter *5*, which is also used for establishing atmospheric pressure in the sample loop of valve *2*. If in the process of displacing gas from syringe *1* the end of the capillary tube inhibits further movement of the plunger, the

2.2 LABORATORY EQUIPMENT FOR GAS CHROMATOGRAPHS 81

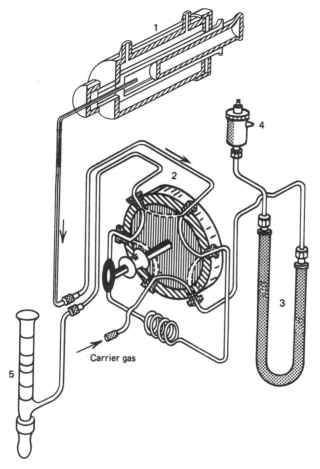

Figure 2.9. System to inject the equilibrium gas from a variable-volume sampling device into the gas chromatographic column. *1*, Variable-volume device; *2*, six-port gas sampling valve; *3*, chromatographic column; *4*, detector; *5*, soap-film gas flow meter.

capillary tube must be removed to an appropriate length. If there is a large amount of liquid in the syringe, the displacement of the gas must be done with syringe *1* in a vertical position in order to avoid the entry of the liquid into the capillary.

In dynamic systems the sampling of an equilibrium gas flow and injection of the gas into a chromatographic column are usually done by a gas

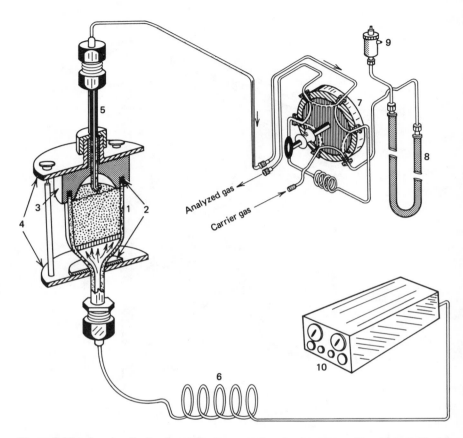

Figure 2.10. Sample collection from a flowing system into the chromatographic column. *1*, No. 16 Glass filter; *2*, rubber washers; *3*, teflon cap; *4*, clamp; *5*, thick-wall glass capillary with a lip; *6*, stainless-steel capillary tube; *7*, gas-sampling valve; *8*, chromatographic column; *9*, detector; *10*, gas flow controlling system.

sampling valve according to the schematic given in Figure 2.10. An apparatus employing a glass filter that allows finely diffused gas to pass through a liquid was used as a container for the extraction of a substance during the process of continuous gas extraction.[10] The velocity and stability of the gas flow is controlled by the pressure/flow regulator *10* of the gas chromatograph and a stainless steel capillary tube *6* used as a dynamic flow resistance. The gas passing from the container continuously flushes the sample loop of

2.2 LABORATORY EQUIPMENT FOR GAS CHROMATOGRAPHS

the gas sampling valve, and the sample is injected into the chromatographic column simply by turning the valve into the sampling position.

An important condition for the successful operation of dynamic systems is the exclusion of fine droplets of liquid from the sample loop of the gas valve. For organic liquids with small surface tension (for example, polyethylene glycols, squalane, dinonyl phthalate), a lip to protect the outlet of the container is sufficient for excluding spray droplets. However, with aqueous solutions it is not sufficient. At the outlet of the container one must place porous filters made of hydrophobic materials (fluoroplastic or forosilicate glass).

Apparatus that first concentrates the impurities contained in the equilibrium gas is used to increase the sensitivity of head-space analysis. The simplest type injects the gas phase by a syringe into an adsorption trap.[11] The concentrate is introduced into the chromatographic column by thermal desorption in the carrier gas flow. Another type uses cryogenic accumulation of the impurities as discussed in the paper by Hurst.[12] Equilibrium distribution of the substance between the phases is done in a fixed-volume flask (Fig. 2.11). The gas space is connected to a test tube for accumulation of the substance and to a 50-ml gastight syringe. To accumulate the impurities, the test tube is lowered into liquid nitrogen, which causes the gas in the flask to begin to condense. This causes a decrease of pressure in the system, which is compensated by compression of the gas volume in the syringe (the plunger of the syringe is lowered). When the volume of the condensed gas reaches 50 ml (the plunger of the syringe is lowered to the limit), the test tube is removed from the liquid nitrogen and the condensed gas begins to evaporate. The pressure in the system increases and the plunger of the syringe begins to rise. The volume of the evaporated gas is brought to 45 ml (approximately 10% of the condensate is left to avoid evaporation of the substances being determined). The process of condensation and evaporation can be repeated many times. When a sufficient quantity of the substance has accumulated, the test tube containing the concentrate is connected by means of a special device to the carrier gas stream of the chromatograph and at an increased temperature the flowing carrier gas transports the trapped substance into the chromatographic column. It must be noted, however, that this original method is useful primarily for qualitative determinations. Quantitative analysis is quite difficult to conduct, especially with repeated condensations.

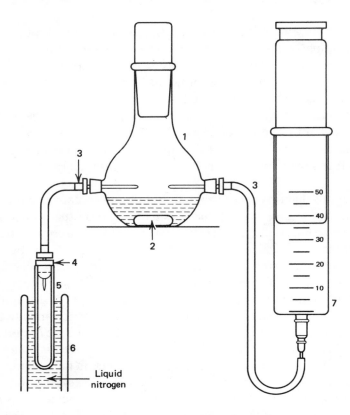

Figure 2.11. System for the cryogenic collection of impurities present in the equilibrium gas. *1*, Flat-bottomed glass flask; *2*, magnetic mixer; *3*, silicone rubber or teflon tubing; *4*, rubber septum; *5*, test tube to collect the sample; *6*, Dewar flask; *7*, gas-tight syringe.

The designs discussed above or similar designs[1] make it possible to increase sensitivity by increasing the amount of equilibrium gas utilized in the analysis. The substance being determined is usually not completely removed from the solution. To completely extract impurities from the solution and transfer them to a concentrator, devices employing dynamic head-space chromatography are used, completely purging the substance from the solution (see, for instance, the description of the attachment to the HP 7675A gas chromatograph in the following section).

2.3 AUTOMATED ACCESSORIES FOR STANDARD GAS CHROMATOGRAPHS AND SPECIALIZED ANALYZERS

The leading manufacturers of gas chromatographs have already developed and are producing specialized automated instrumentation for head-space analysis. These systems usually utilize standard pharmaceutical vials as the containers in which equilibrium is established.

The accessory Model HS-250 for the Carlo Erba universal gas chromatograph[13] (Fig. 2.12) consists of a sampling unit for the equilibrium gas that utilizes a syringe, and a control unit. The sampling unit consists of a metal turntable with recesses for glass vials of 5–10-ml capacities and a holder for a gas syringe. These are temperature-controlled in the range of 35–100°C with an accuracy of ±0.5°C. The turntable can accommodate up to 40 vials each closed with rubber caps. The surface of the turntable is coated with Teflon. The sample volume can be varied in the range from 0.1 to 2 ml by regulating the length of the plunger movement of the gas syringe. In the sampling mode, the temperature, the time of equilibration, flushing with air, filling the gas syringe with the sample and introducing the sample into the analytical column, and the time intervals between samplings are set on the control unit and the entire operation is carried out and controlled automatically.

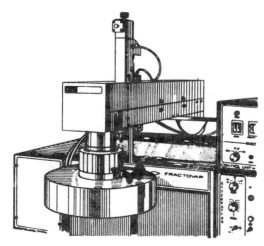

Figure 2.12. HS-250 accessory for automatic head-space analysis of the Carlo Erba Co.

The determination of volatile substances in polymers by means of direct sampling of the gas phase above a solid sample (without using solvents) requires operation at temperatures above 100°C. For this purpose automated samplers have been constructed with oil, air, or metal thermostats.

A sampler with an air thermostat[14] may operate at temperatures up to 300°C, heating simultaneously thirty-six 20-ml vials containing the samples. Although an air thermostat requires more time for equilibration,[15] it is claimed that air is preferable to oil in the thermostat because in this way the possibility of contaminating the samples is excluded. The automatic analyzer with such a sampler gave good results for the determination of residual solvents in commercial polymeric substances.[15]

The Model HS-6 accessory for head-space analysis, which is based on the pneumatic method of sampling the equilibrium gas from glass vials closed with a rubber cap, has been developed by Perkin-Elmer for the Sigma series of gas chromatographs.[16] This unit (Fig. 2.13) consists of a rotating temperature-controlled magazine with six recesses for 6-ml glass vials and a control unit for programming the time and temperature of the sampling. The magazine is mounted on the panel of the injector of the chromatograph. The gas flow system of the apparatus with the HS-6 (Fig. 2.14) includes an optional precolumn, an injector with a needle for puncturing the rubber cap on the sample vial, and two valves with an electric control. The carrier gas can also pass through valve 7 directly into the analytical column. The

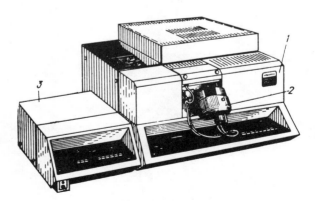

Figure 2.13. General view of a Perkin-Elmer Sigma series gas chromatograph with the HS-6 head-space sampling accessory. *1*, Gas chromatograph; *2*, temperature-controlled sample-holder magazine; *3*, control unit of the head-space sampling accessory.

2.3 AUTOMATED ACCESSORIES AND SPECIALIZED ANALYZERS 87

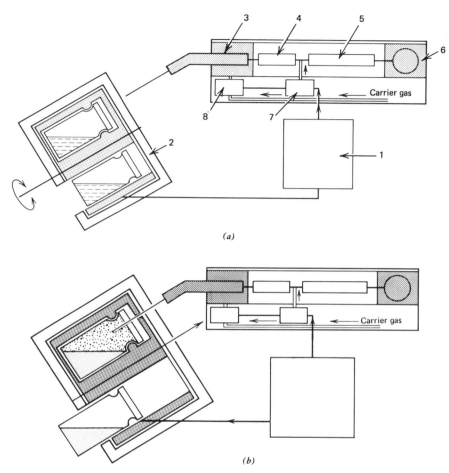

Figure 2.14. Schematic of the gas chromatograph with the HS-6 head-space sampler. (*a*) positioning of the sample vial; (*b*) sampling. *1*, Control unit of the head-space sampling accessory; *2*, temperature controlled magazine to hold the vials; *3*, injector of the gas chromatograph; *4*, precolumn; *5*, chromatographic column; *6*, detector; *7*, *8*, electrically controlled gas valves.

magazine construction allows transfer of the sample vials along the axis to one of two positions. In the lower position (Fig. 2.14*a*), a new vial containing a sample can be introduced by turning the magazine to one of six fixed positions. In the upper position of the magazine (Fig. 2.14*b*) the injector needle punctures the rubber cap and enters the gas space above the sample.

The analyzed liquid or solid samples are placed into glass vials which are hermetically sealed by rubber caps with an aluminum crimp on the neck of the vial. The vials to be analyzed are placed in the recesses of the magazine (Fig. 2.14a) and are kept in the lower position until constant temperature and equilibrium distribution between the phases is established. During the preparatory period the carrier gas passes directly through the analytical column. Sampling is done by moving the magazine into the upper position (Fig. 2.14b). The injector needle enters the gas space of the vial, valve 7 closes, and valve 8 opens. Now the carrier gas enters the gas phase of the vial and the same pressure will develop in the vial as at the chromatographic column inlet. Upon ending this process, valve 8 is closed (valve 7 remains closed), and the carrier gas flow is temporarily interrupted. Now an aliquot of the gas phase from the vial enters the analytical column. The timing of gas sampling from the vial (5 sec) is regulated by an electronic timer with high accuracy and serves as the beginning of the chromatographic analysis. After sample introduction, valve 7 opens, and the original initial pressure is reestablished in the gas system of the chromatograph by the carrier gas.

If the sample gas contains the vapors of high-boiling substances for which analysis is not required, the control unit can be programmed to backflush the precolumn by opening valve 8, closing valve 7, and removing the injector needle from the flask.

An example of the devices manufactured for head-space analysis utilizing complete gas extraction of solutions is the HP 7675A, the accessory manufactured for Hewlett-Packard gas chromatographs.[17] The attachment represents an individual unit (Fig. 2.15) containing controls and programming of temperature and time, and a gas sampling device (Fig. 2.16) that includes temperature-controlled gas valves and a trap with an adsorbent.

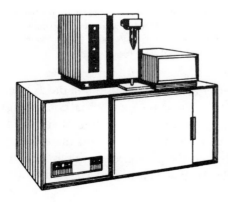

Figure 2.15. General view of the HP-7675A head-space accessory (top unit) mounted on a Hewlett-Packard model 5840A gas chromatograph.

2.3 AUTOMATED ACCESSORIES AND SPECIALIZED ANALYZERS 89

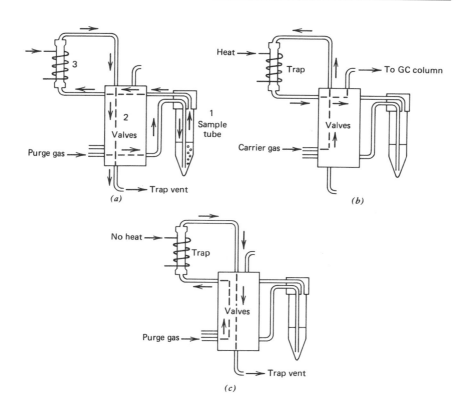

Figure 2.16. Flow schematic of the HP-7675A accessory during the various operation cycles. (a) *Purge cycle.* The sample to be investigated is removed from its solution and is trapped on the adsorbent. (b) *Desorb cycle.* The sample collected in the trap is thermally desorbed into the carrier gas flow. (c) *Trap-cleaning cycle.* The system is backflushed to remove the heavy components from the trap. *1*, Extractor containing the sample; *2*, thermostat and valves; *3*, trap with the adsorbent.

The flask containing the sample for extraction of the volatile components by a clean air flow is mounted on the front panel of the instrument and is at room temperature. Up to 50 ml of a sample solution can be analyzed.

Carrying out an analysis consists of three cycles. In the purge cycle (Fig. 2.16a) the volatile components to be analyzed are purged from the solution by a gas and trapped in a cooled tube containing the adsorbent. In the desorb cycle (Fig. 2.16b) the adsorption tube is heated quickly and connected to the carrier gas flow, which transfers the desorbed substances into the chromatographic column. In the trap-cleaning cycle (Fig. 2.16c), the heated adsorption tube is backflushed to clean the system of the remaining "heavy" components. All operations are automated. The duration of each cycle is regulated in the interval of 0–99 min after 1 min. The temperature of the switching gas valves is kept constant at 50, 90, 120, 150, 200, 250, or 300°C, and the adsorption trap is heated 180°/min up to any temperature in the range of 100–350° in steps of 50°. It is also possible to operate the HP 7675A with a mass spectrometer.

An universal accessory for the injection of equilibrium vapor into a chromatographic column was recently recommended by Baylis and co-workers.[18] The operation principle of this design is based on continuous gas extraction, that is, on passing a carrier gas flow gas above or through a temperature-controlled solution and transferring the vapors of volatile substances into a chromatographic column cooled to −40°C. The instrument is mounted on the injection block and is supplied with a reservoir for the sample solution, heating elements and with a system for switching the gas flow directly to an analytical column either through a solution or by-passing it. The moment the carrier gas flow is switched directly to the chromatographic column is the end of sample injection and simultaneously the beginning of chromatographic analysis and is usually accompanied by an increase in the oven temperature. The basic advantage of this design is that it permits the injection of large quantities of an equilibrium gas (up to several liters), into the chromatograph, which secures high analytical sensitivity. When this attachment is used for quantitative determinations, it is necessary to take into consideration the regularities of continuous gas extraction, the basic relationships of which are given in Chapter 1 (Section 1.4).

Specially constructed Perkin-Elmer automated instruments are widely used in head-space analysis. The company has been developing three such analyzers, the models F-40,[19] F-42,[20] and F-45.[21] These instruments are universal gas chromatographs supplied with systems for temperature control of the containers to establish an equilibrium and electropneumatic sampling of the equilibrium gas directly into a chromatographic column. The development of three consecutive models resulted in continuously improved temperature control and sampling system, and general improvements in the design of the gas chromatograph. Therefore, here only the newest model is discussed.[19-25]

The model F-45 head-space analyzer (Fig. 2.17) is an up-to-date gas chromatograph with a differential gas system, a temperature-programmed packed or capillary chromatographic column, and five of the most popular

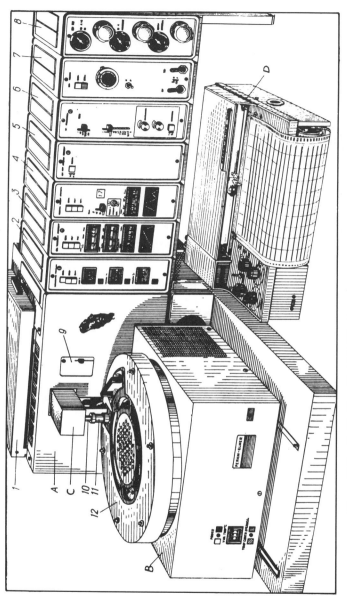

Figure 2.17. Perkin-Elmer Model F-45 head-space analyzer. (*A*) Universal gas chromatograph; (*B*) constant-temperature thermostat for the sample; (*C*) automatic sampler; (*D*) recorder. *1*, Chromatographic column oven; *2*, control unit for the column oven and detectors; *3*, temperature programming unit for the column oven; *4*, programmer of the head-space sampling system; *5*, power supply unit of the head-space sampling system; *6*, amplifier, power supply and control unit for the flame-ionization detectors; *7*, power supply and control unit for the thermionic detector; *8*, gas supply unit for the flame-ionization detector; *9*, temperature control of the sampling needle; *10*, directional plunger; *11*, movable cylinder; *12*, turntable with places for the sample vials.

detectors: two universal—differential flame ionization detector and katharometer—and three selective—electron-capture (for halogen-containing substances), flame photometric S and P containing substances), and thermionic N and P containing substances). Simultaneous operation of two ionization detectors is possible. The gas system is designed to permit the backflushing of the chromatographic column for removing slightly volatile substances and for rapid preparation of the instrument for the next analysis. There is an evaporator for liquid samples that allows the instrument to be used not only for head-space analysis but also as a common universal gas chromatograph.

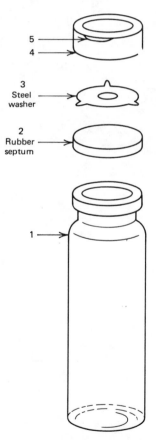

Figure 2.18. Sample vial used in the F-45 head-space analyzer. *1*, Glass vial; *2*, rubber septum; *3*, steel washer; *4*, aluminum cap; *5*, groove for releasing the excess pressure;

2.3 AUTOMATED ACCESSORIES AND SPECIALIZED ANALYZERS

To establish an equilibrium, glass vials are used that have a capacity of 24 ml and a rubber septum that is sealed to the vial with an aluminum cap (Fig. 2.18). To secure complete sealing, a star-shaped steel washer *3* is placed between the rubber septum *2* and the aluminum cap *4*. This keeps the container hermetically sealed at increased pressure over a prolonged period (up to several days). In the aluminum cap there is a special groove *5* to release the pressure in the container when it reaches a dangerous level (approximately 6 atm). After reduction of the pressure the hermetic seal is reestablished.

Depending on the nature of the analyzed object and the conditions of conducting the analysis, four types of rubber septa are used:

1. Butyl rubber septa, the cheapest, are durable up to 120°C, but they absorb nonpolar substances, such as hydrocarbons. Polar compounds as well as alcohols are absorbed only insignificantly.
2. Butyl rubber septa coated on one side with Teflon are also durable up to 120°C in corrosive media. They are reasonably cheap and do not absorb either polar or nonpolar substances. However, once they are punctured they lose their inert properties due to disturbance of the protective Teflon layer. Repeated puncturing is not recommended.
3. Silicone rubber septa with a Teflon layer are the most expensive and have the same properties as the Teflon-coated butyl rubber septa but withstand temperatures up to 190°C.
4. Silicone rubber septa with an aluminum coating are cheaper than those coated with Teflon and withstand heat up to 150°C. They are widely used at high temperatures. Repeated puncturing is not recommended, nor is use with samples that contain strong acids such as HCl.

The system of preparing the samples for head-space analysis and sampling the equilibrium gas consists of a liquid thermostat filled with silicone oil into which a round turntable with 30 positions for the sample vials is lowered, and a device for the pneumatic sampling of the equilibrium gas which is automatically controlled. The turntable of the thermostat can rotate around a vertical axis into any of 30 fixed positions to change a sample, or in the process of sampling can move along this axis into middle or upper positions. The samples in the vials do not come into contact with the silicone oil because the oil level does not reach the upper surface of the

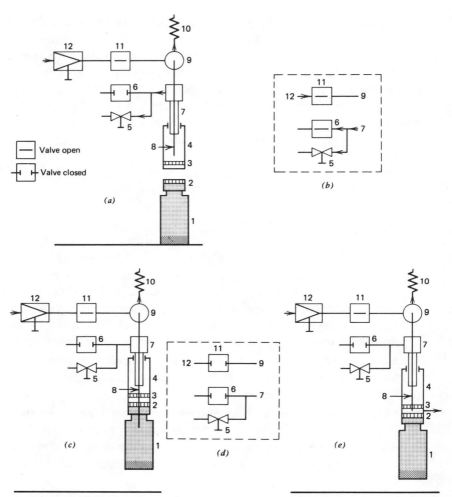

Figure 2.19. Operation schematic of the pneumatic sampling system of the Perkin-Elmer automatic head-space analyzers. For the explanation of the various stages of its operation, see the text. *1*, Glass vial with the sample; *2, 3*, rubber septum; *4*, movable cylinder; *5*, needle valve; *6*, gas valve; *7*, directional plunger of the movable cylinder; *8*, sampling needle; *9*, inlet to the chromatographic column; *10*, chromatographic column; *11*, gas valve, *12*, pressure (flow) regulator.

2.3 AUTOMATED ACCESSORIES AND SPECIALIZED ANALYZERS 95

turntable even when the unit is in the lower position. The thermostat temperature is kept within the range of 35–150°C with an accuracy of ±0.1°C. In previous models, water thermostats were used and sample temperatures could not exceed 100°C.

The schematic of pneumatic sampling of the equilibrium gas is given in Figure 2.19. A sample is injected by means of the sampling capillary *8*, which is rigidly fastened in the hollow controlling plunger *7* and connects the inside space of the movable cylinder *4* with point *9* of the gas inlet of the chromatograph. The upper part of cylinder *4* is hermetically connected with plunger *7* along which it can move. The lower part of cylinder *4* is hermetically sealed with a rubber membrane *3* and has a release port through which gas can exit into the atmosphere.

The pressure of the carrier gas at point *9* is controlled by regulator *12*. The electrically controlled gas valve *11* allows to temporary disconnect the carrier gas flow to the chromatographic column. Gas valve *6*, which is also controlled electrically and needle valve *5* control purging of the inside space of cylinder *4*.

The sampling of an equilibrium gas from a sample vial is done in the following manner: The turntable of the liquid thermostat that is in the lower position is turned to the position where sample vial *1* would be directly beneath cylinder *4* of the pneumatic sampler. Meanwhile, valve *11* is open and the carrier gas passes into the chromatographic column (Fig. 2.19*a*). Then, by command from an electronic regulator, the turntable with the sample vials begins to rise into the upper position. In the process of rising, valve *6* opens to have the carrier gas purge cylinder *4* (Fig. 2.19*b*). At the moment the vial reaches the upper position, valve *6* closes and sampling needle *8* punctures membrane *3* and septum *2* and enters the gas space of the sample vial (Fig. 2.19*c*). Because of the pressure difference, the carrier gas begins to pass through the needle *8* into the sample vial. At this moment a light on the programmer (Fig. 2.17) indicates the number of the sample vial and records it.

When the pressures at point *9* of the gas sampling device and at flask *1* are equalized, valve *11* is closed (Fig. 2.19*d*). Now the carrier gas flow is disconnected, and the pressure at point *9* begins to drop. The gas from the vial enters the chromatographic column through sampling capillary *8* due to the developed pressure drop. The volume of the sample introduced depends on the period how long the carrier gas flow was interrupted, that is, the duration of the shutoff of valve *11*.

The sampling process is stopped by opening valve *11*, because the pressure at point *9* quickly reestablishes to a pressure higher than in the sample vial. After stopping the introduction of the sample into the column, the vial remains for several seconds in the upper position. Then the turntable of the liquid thermostat is moved into the lower position, and the sampling system acquires the position indicated in Fig. 2.19*a*.

At the end of the chromatographic analysis the flask is raised into the middle position (Fig. 2.19*e*). In the process of rising, valve *6* opens to purge cylinder *4*. When the flask reaches the middle position, valve *6* closes, and the needle *8* is now between membrane *3* and septum *2*. In this position the sampling needle is purged to remove the substances remaining from the previous sample. After cleansing the needle, the turntable unit with the vials moves to the lower position and at this moment the analysis of the next sample vial can start.

The length of individual sampling cycles is controlled on the programming unit. Sampling reproducibility is better than 1%.

The automatic Perkin-Elmer head-space analyzers are widely used to control the content of toxic substances in biological samples (mainly in blood and urine), to determine volatile substances in polymer materials, and for the analysis of environmental media (air and water) and food products.

REFERENCES

1. H. Hachenberg and A. P. Schmidt, Gas Chromatographic Head-space Analysis, Heyden, London, 1977, 125 pp.
2. H. Zenz and H. Klaushofer, *Mitt. Vers. Anst. Gär. Gew.*, **22**, 175 (1968) (cited in ref. 1).
3. H. Pauschmann, *Chromatographia*, **3**, 376 (1970).
4. G. Göke, *Chromatographia*, **5**, 622 (1972).
5. D. Jentzsch, TH. Krüger, G. Lebrecht, G. Dencks, and J. Gut, *Z. Anal. Chem.*, **236**, 112 (1968).
6. A. G. Vitenberg, I. L. Butaeva, and Z. St. Dimitrova, *Zavodsk. Lab.*, **8**, 931 (1975).
7. A. G. Vitenberg, I. L. Butaeva, and Z. St. Dimitrova, *Chromatographia*, **8**, 693 (1975).
8. W. Piorr, *Glass-Instr.-Tech.*, **12** 175 (1968).

REFERENCES

9. P. Ronkainen, *Kem. Teollismus*, **26**, 215 (1969) (cited in ref. 1).
10. A. G. Vitenberg and M. I. Kostkina, *Zh. Analit. Khim.*, **34** 1800 (1979).
11. M. Gottauf, *Z. Anal. Chem.*, **218**, 175 (1966).
12. R. E. Hurst, *Analyst*, **99**, 302 (1974)
13. "A New Accessory for the Determination of Volatile Substances by Head-space Gas Chromatography. Automatic Sampler Model HS 250," Publication DT HS 250E-AGG-L2-12-76, Carlo Erba, Milano, 1976.
14. R. E. Arner, D. E. White, J. B. Paush, and G. F. Pfeiffer, Paper No. 278, Pittsburgh Conference on Analytical Chemistry and Applied Spectroscopy, Cleveland, Ohio, 1978.
15. R. P. Lattimer, J. J. Cangemi, and J. B. Pauch, Paper No. 279, Pittsburgh Conference on Analytical Chemistry and Applied Spectroscopy, Cleveland, Ohio, 1978.
16. "Head-Space Sampler HS-6 for Gas Chromatography." Publication 1738/1.78 Bodenseewerk Perkin-Elmer & Co. Überlingen; B. Kolb, P. Pospisil, M. Jaklin, and D. Boege, *Chromatogr. Newslett.*, **7**, 1 (1979).
17. Hewlett-Packard 7675A Purge and Trap Sampler.
18. M. A. Baylis, J. D. Green, and S. R. Massey, *Chem. Ind.*, **1979**, 353.
19. "Automatic GC Multifract F-40 for Head-Space Analysis," Publication F-256/811E Bodenseewerk Perkin-Elmer & Co. Überlingen.
20. "Model F-42 Head-Space Analyzer," Publication 1377/4.7 Bodenseewerk Perkin-Elmer & Co. Überlingen.
21. "Gas Chromatography Model F-45 Head-Space Analyzer," Publication 1809/9.78 Bodenseewerk Perkin-Elmer & Co. Überlingen.
22. D. Jentzsch, H. Krüger, and G. Lebrecht, *Angew. Gas-Chromatogr.*, Vol. 10E, Bodenseewerk Perkin-Elmer GmbH, Überlingen, 1967, 21 pp.
23. B. Kolb, E. Wiedeking, and B. Kempken, *Angew. Gas-Chromatogr.*, Vol. 11E, Bodenseewerk Perkin-Elmer GmbH, Überlingen, 1968, 9 pp.
24. B. Kolb, *Angew. Gas-Chromatogr.*, Vol. 15E, Bodenseewerk Perkin-Elmer GmbH, Überlingen, 1972, 12 pp.
25. B. Kolb, *J. Chromatogr.*, **122**, 553 (1976).

CHAPTER THREE

Application of Head-Space Analysis for the Quantitative Determination of Impurities

3.1 ANALYSIS OF WATER AND AQUEOUS SOLUTIONS

One of the most important applications of quantitative head-space analysis is the determination of trace organic substances in drinking, natural, and industrial waters and solutions and in sewage. Here all the basic techniques of head-space analysis are applicable. The analytical technique depends upon the sensitivity and accuracy required and on the nature of the analyzed components. Trace substances with small distribution coefficients (water–air, $K < 10$) can be determined by direct gas-chromatographic analysis of the equilibrium vapor. These analyses use universal detectors. (The flame ionization detector is especially convenient due to its insensitivity toward water.) Hydrocarbons, halogen and sulfur derivatives present in trace quantities in water have such small values of K. The majority of investigations on the head-space analysis of water are devoted to the determination of these impurities, mentioned in Chapter 1.

The limit in the determination of volatile impurities by the direct analysis of the vapor phase can vary considerably. These limits are variable because of differences in the distribution coefficients and also because of differing detector sensitivity with respect to various compounds. For example, with a Perkin-Elmer F-45 analyzer with a flame-ionization detector, trichloroethylene in concentrations as low as $6 \times 10^{-6}\%$ can be determined in the vapor of an aqueous solution heated to 80°C (10 ml). The limit of detecting dibromochloromethane is only $2.33 \times 10^{-4}\%$, that is, the sensitivity is 40 times less.[1] The detection limit for polyhalogen derivatives can be considerably lowered by the use of electron-capture detectors.* However, for the determination of sulfur and phosphorus compounds it is expedient in many cases to use a flame photometric detector. With direct head-space analysis of many waters and aqueous solutions it is possible to influence the distribution coefficients of the volatile impurities present in considerable and widely varying concentrations. Frequently, the distribution coefficients change substantially due to hydrogen ion or mineral salt concentrations that are not controlled gas-chromatographically. The influence of these factors may be included or excluded in several ways. One method consists in adding a buffer solution to the analyzed sample. The sample solution has a fixed pH and therefore excludes the effect of fluctuations in acidity or alkalinity. The

---

*The electron-capture detector requires the complete separation of water vapors.

3.1 ANALYSIS OF WATER AND AQUEOUS SOLUTIONS

influence of varying concentrations of salts can be suppressed by adding an excessive amount of salt to the solution. This allows a leveling of the fluctuations of the salt concentration in samples.

These methods are used during the head-space analysis of aqueous solutions and the waste waters of sulfate pulp industries in determining the content of sulfur compounds: hydrogen sulfide, methylmercaptan, ethylmercaptan, dimethyl sulfide, and diethyl disulfide.[2-5] A water sample (10-20 ml) is collected using a glass thermostated syringe (see Chapter 2). The sample is mixed with an equal volume of KCl-HCl buffer solution of pH 2 saturated with sodium sulfate. This mixture suppresses the dissociation of hydrogen sulfide and lowers and stabilizes the values of the distribution coefficients of all sulfur-containing compounds. Furthermore, an approximately twofold increase in sensitivity is achieved. The equilibrium gas above the solution is introduced into the chromatographic column by being pushed through the loop of the gas sampling valve using the plunger of the syringe. Using flame ionization or micro coulometric detector and samples of 0.3-0.8 ml, the sulfur compounds can be determined in industrial discharges at a level of 10^{-7}%. These results can be determined with an error up to 20%. With concentrations at the order of 10^{-4}%, the error does not exceed 12%. The coulometric detector allows a high selectivity in the determination of sulfur compounds contained as impurities with a background of hydrocarbon and oxygen compounds.

The other possible way of tracking the fluctuations in the distribution coefficients consists of measuring them directly in the analyzed sample. This method is used in head-space analysis of aromatic hydrocarbons in water.

The importance of determining the complex composition of industrial waste water lies in the relatively high toxicity of aromatic hydrocarbons. On the other hand, there is the possibility of rapid and reliable detection of the simplest aromatic hydrocarbons in stratal waters. These, on the order of 0.05 mg/liter or greater, recently became of interest to geochemists in relation to the investigation of oil deposits. Benzene and its closest homologs are characterized by a relatively good (for hydrocarbons) solubility in water. Therefore, they are present in the stratal waters contacting with oil deposits in quantities much greater than other hydrocarbons. The presence of the simplest aromatic hydrocarbons in stratal waters is presently being considered as an important, direct, and effective indicator for the discovery of oil and gas condensate deposits. McAuliffe[6,7] was the first to indicate the expediency of applying head-space analysis for this purpose. Stratal water

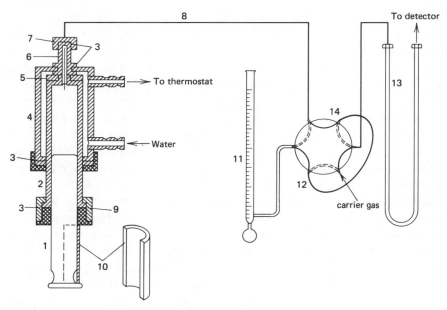

Figure 3.1. System to introduce the water vapor for the determination of trace aromatic hydrocarbons present. *1*, Plunger; *2*, 100-ml syringe; *3*, rubber seals; *4*, poly(methyl methacrylate) thermostating jacket; *5*, epoxy seal; *6*, teflon sleeve; *7*, union nut; *8*, stainless steel capillary tube, 1-mm i.d.; *9*, adjustable ring to regulate the plunger and secure its position; *10*, holder; *11*, manometer; *12*, thermostated sample loop; *13*, chromatographic column; *14*, six-port valve.

can contain various amounts of mineral salts, which considerably influence the values of the distribution coefficients. The most effective application in this case is that of repeated gas extraction. As was shown in Chapter 1, the chromatographic analysis of the equilibrium vapor phase before and after its substitution by a fresh portion of gas allows both the measurement of the distribution coefficient for a given sample and the determination of its concentration to be combined in one experiment.

An improved method of head-space analysis of natural and sewage waters for benzene and toluene[8,9] content assumes the use of a sampling system for the equilibrium vapor as shown in Figure 3.1. In it, 5–50 ml of the investigated water is introduced by means of a calibrated medical syringe into the thermostated syringe *2* with a volume of approximately 100 ml. The sample is shaken for 15–30 min at a constant temperature (a little below room temperature*). The syringe is connected to a heated six-way valve *14* by

---

*To exclude condensation of the vapors.

stainless steel capillary *8*. The sample loop *12* of this system has a volume of approximately 1.5 ml. Upon raising plunger *1*, the sample loop, which is heated to 110–120°, is filled with the gas phase. Twenty seconds after atmospheric pressure is achieved in the system, valve *14* is turned into the sampling position. A sample of the gas phase is introduced into the chromatographic column. Then the gas phase is completely removed from the syringe and an equal volume of pure air is collected, which returns plunger *1* exactly to the previous position (according to gauge *11*). Again the syringe is shaken until equilibrium is established, held for 15–30 min, and the procedure is repeated. Figure 3.2 shows typical chromatograms obtained in such a manner for samples of subsurface waters near oil deposits. The content of

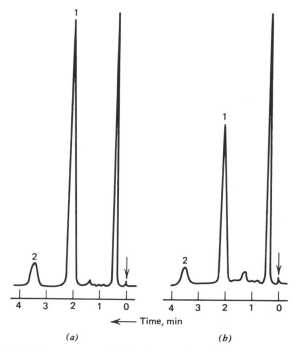

Figure 3.2. Typical chromatograms of the head-space of a subsurface water sample taken near oil deposits. (*a*) The first gas sample taken from the head-space of the equilibrium container; (*b*) The second gas sample taken from the head-space of the equilibrium container. *1*, Benzene; *2*, toluene. Perkin-Elmer Model 900 gas chromatograph equipped with flame ionization detector. Column: 2 m × 3 mm i.d., containing 15% polyethylene glycol adipate on Dinokhrome H (0.25–0.32 mm fraction). Column temperature: 70°C. Carrier gas (nitrogen) flow rate: 30 ml/min.

benzene or toluene (C_L^0) in them can be calculated with the help of equation

$$C_L^0 = \frac{m}{v}\left(\frac{A}{A-A'}\right)\left(\frac{V_G}{V_L}\right) \tag{3.1}$$

where V_L is the volume of the water sample, V_G is the volume of the equilibrium gas phase in syringe 2, v is the volume of the sample loop 12, A and A' are the areas of the peaks on the chromatograms (the integrator readings) before and after phase substitution, and m is the quantity of benzene (or toluene) in a sample introduced into the chromatograph. This value is obtained from a previously constructed calibration curve $m = f(A)$ for a given sample loop.

The duration of analysis is approximately 1.5 hr. The reproducibility corresponds to a relative standard deviation of 0.05–0.1, with concentrations of benzene and toluene in the range of 0.001–45 mg/liter.

The other case of direct head-space analysis for substances of unknown distribution coefficients (the addition of known quantities of the analyzed substances as standards) was recommended for the determination of hydrocarbons in water by Drozd, Novák, and Rijks.[10] The addition of known quantities of paraffinic and aromatic hydrocarbons was achieved by the introduction of 1 μl of standard 0.1% acetone solutions into a closed thermostated 100-ml container containing 50 ml of the analyzed water. With vapor samples of 1 ml and working with capillary columns without splitting the sample, the standard mixtures containing of 0.2–26 μg were determined with an average error of less than 10%, approximately the same as in the above-described method of repeated gas extraction.

The direct gas-chromatographic analysis of the vapors of dilute aqueous solutions using the flame-ionization detector allows the determination of volatile hydrocarbons in concentrations as little as 10^{-6}–10^{-7}%. To further decrease the limit of detection requires the use of preliminary concentration. Gottauf[11] described a simple concentration technique useful for quantitative determinations in head-space analysis that allows an increase in sensitivity to 10^{-8}%. A 10-ml sample of the analyzed aqueous solution is placed in a glass syringe vertically fixed at the 70-ml mark, with a total volume of 100 ml. The sample is first purged with pure helium. Into the same syringe 12 ml of a salting-out reagent (Table 3.1) is then introduced. The opening is closed with a glass stopper, and the mixture is shaken for 10 min. Then the stopper is removed and replaced by the cryogenic trap (i.d. 1 mm and length 50 cm) made from a stainless steel capillary. The ends of

3.1 ANALYSIS OF WATER AND AQUEOUS SOLUTIONS

Table 3.1 Salting-Out Reagents for the Head-Space Analysis of Volatile Substances in Aqueous Solutions[a]

Reagent	Quantity of Dissolved Substances in 100 ml of Water
Neutral solution of sodium sulfate	45 g NA_2SO_4
	1.42 g $NaHPO_4 \cdot 2\ H_2O$
	0.71 g KH_2PO_4
Acidic solution of sodium sulfate	45 g NA_2SO_4
	0.12 g H_2SO_4
Alkaline solution of sodium sulfate	45 g NA_2SO_4
	0.4 g NaOH
Acidic solution of sodium sulfate	45 g Na_2SO_4
	0.28 g $Na_2HPO_4 \cdot 2\ H_2O$
	1.6 g KH_2PO_4
	0.02 g $(NH_2OH)_2 \cdot H_2SO_4$
Solution of sodium hydroxide	100 g NaOH

[a] The indicated quantities of reagents are dissolved in water and heated to 40°C, and for 2 hr the solutions are purged with 12 liters of pure nitrogen to remove the volatile impurities.

Source: M. Gottauf, ref. 11.

the capillary are equipped with soldered brass fittings to which rubber tubes are connected, as shown in Figure 3.3. The middle part of the capillary (5 cm in length) is filled with a chromatographic column packing and is

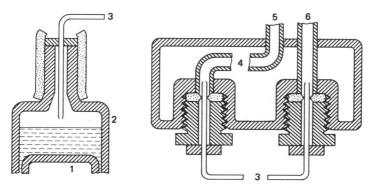

Figure 3.3. System to connect a cryogenic trap to a syringe containing the analyzed solution (left) and to the gas chromatograph (right). *1*, Syringe plunger; *2*, syringe; *3*, ends of the cryogenic trap consisting of a 50 cm × 1 mm i.d. capillary tube; *4*, heated tube supplying the carrier gas; *5*, connection to the carrier gas inlet; *6*, connection to the chromatographic column.

lowered into a dewar flask containing liquid air. By the motion of the plunger, the gas phase is pulled through the trap. After concentration in the trap, the system is connected to the gas chromatograph (Fig. 3.3, *right*) and lowered into boiling water. Quantitative determinations require calibration by solutions of known concentration, which are used immediately following preparation. This technique allows the determination of trace aromatic substances in aqueous foodstuffs. The relative standard deviation was found to be ±0.05%.

Hydrocarbon impurities in water have been determined at the level of parts per billion (ppb).[10] The components of the vapor phase were concentrated in the capillary with a film of OV-101 methyl silicone oil and cooled to −50–60°C before introduction into the column and then quickly heated to 150°C.

The described methods of concentration of the vapor phase do not assume the total extraction of volatile components. The fraction remaining in the liquid phase can be taken into consideration in calculations, but in the case of substances with large distribution coefficients and for the analysis of trace impurities the sensitivity of the determination may be insufficient. Modern methods of water analysis at concentration levels on the order of micrograms per liter allow the possibility of total extraction of the volatile impurities by means of gas extraction, that is, "stripping" by means of special devices. The stripping of water samples can be achieved with or without subsequent cryogenic or sorption concentration if the conditions of stripping secure sufficiently high concentrations of the detected components in the vapor phase. An important factor in the accumulation of volatile substances in the vapor phase is the increase of the temperature of the solution and the instrumentation for stripping. As a rule, stripping assumes the heating of the analyzed sample. Figure 3.4 shows a stripping device for water samples with direct introduction of the vapor phase into a chromatograph. A sample of water (2–10 ml) introduced into the upper part of the device is pushed by a flow of carrier gas through four heated chambers flowing along their walls as a thin film that aids in the quick establishment of equilibrium. The exchange conditions between the phases in the device with four chambers approximately corresponds to a fourfold consecutive gas extraction. The impurities converted into the vapor phase are purged by the carrier gas into the chromatographic column. The inner volume of the device should not exceed the maximum allowable volume of the gas sample for a given chromatographic column. The degree of extracting the majority

3.1 ANALYSIS OF WATER AND AQUEOUS SOLUTIONS

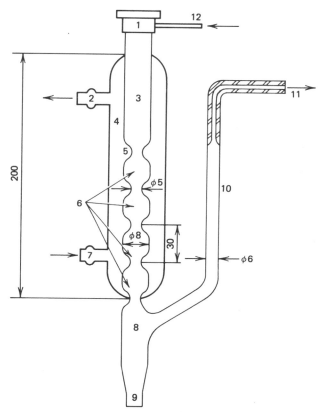

Figure 3.4. Device for stripping water samples with direct introduction of the vapor phase into the gas chromatograph. Dimensions given are in mm. *1*, Inlet for the water sample; *2* and *7*, connections for the thermostating liquid; *3*, sample collector; *4*, thermostating jacket; *5*, capillary narrowing; *6*, chambers; *8*, collector; *9*, discharge; *10*, air cooler; *11*, carrier gas exit; *12*, carrier gas inlet.

of hydrocarbons and halogen derivatives under such conditions is 80–98%. The minimum concentration determined (in an aqueous sample of 10 ml) is 10^{-7}%.[12] It is possible to work with aqueous solutions as well as with homogenized biological materials.

If the temperature of the liquid during stripping increases to the boiling point, then gas extraction is not necessary, as its role is performed by the water vapor during boiling. The combination of distillation with head-space analysis of the distillate is a modification of stripping.[13] Distillation as a method of stripping is effective in the analysis of traces of volatile polar organic compounds—low molecular weight alcohols, ketones, and aldehydes—which are poorly separated by liquid extraction and sorption from aqueous solutions. The method of analysis for recycled sewage waters of military hospitals described by Chian et al.[13] includes the distillation of 100 ml of water using a 20-cm fractionating column and sampling the first 1.5 ml of the distillate. One milliliter of the distillate is placed into a standard container of 15-ml volume, saturated with sodium sulfate and then subjected to head-space analysis. The peak areas of the simplest alcohols, ketones, and ethyl acetate are proportional to their concentration in the initial water sample. The proportionality coefficients are established by the analysis of standard mixtures under similar conditions with concentrations from 8×10^{-7}% to 1.6×10^{-3}%. The detection limit for methyl and ethyl alcohol was 8×10^{-7}%. For their homologs and ketones, it was 4×10^{-7}%.

The separation of volatile impurities using the vapor of boiling water is involved in several methods using special instrumentation.[14-16] The distinguishing feature of such methods is the direct passage of the jet of hot vapor into the chromatographic column where the separation of impurities takes place. The overheated water vapor functions as the carrier gas. Certain modern chromatographs employ such devices to extract the gases and volatile impurities from the solutions. For example, the accessory PFO-49 for the Model Tswett 100 gas chromatographs manufactured in the U.S.S.R. works on the principles of frontal enrichment and allows a two- to threefold increase in the sensitivity of the analysis.

The "chromadistillation" of Zhukhovitskii and co-workers* should be considered as a modification of stripping in which volatile impurities from a sample of a solution are isolated by repeated evaporation and condensation. This process is carried out in a special column packed with metal or glass beads and placed before the chromatographic column.

Much more popular are types using inert gases to strip the solutions coupled with a supplemental cryogenic and sorption concentrator. Such a system assures maximum sensitivity in the investigation of substances important to environmental protection (less than 1 mg of volatile com-

---

*There are many publications by these authors, but most of them are concerned with the theory of chromadistillation as a method for the separation of components of liquid mixtures. The problems of the enrichment and analysis of volatile components are considered in the paper by A. A. Zhukhovitskii, B. P. Okhotnikov, S. M. Yanovskii, and L. G. Novikova, *Zh. Anal. Khim.*, **34**, 545 (1979). See also A. A. Zhukhovitskii, S. M. Yanovskii, and V. P. Shvarzman, *J. Chromatogr.*, **119**, 591 (1976).

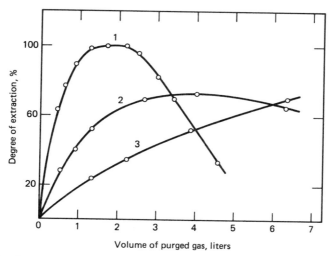

Figure 3.5. Degree of extraction of hydrocarbons upon stripping a 100-ml water sample containing ppm-concentrations of the hydrocarbons and trapping them on an adsorbent (0.4 g). Desorption temperature; 170°C. *1*, Benzene; *2*, n-heptylbenzene; *3*, 2-methylnaphthalene.

ponents per metric ton of water). The complete extraction of volatile components when stripping is combined with adsorption and subsequent desorption depends upon the successful completion of each of these three processes. Yet the optimum conditions in the successive stages of stripping and desorption usually do not correspond to the optimum conditions of the sorption concentrator. Also, the conditions required for complete extraction of various impurities can considerably differ, even within the confines of one class of compounds. Figure 3.5 illustrates the degree of extraction of three aromatic hydrocarbons by stripping from 10^{-4}% solutions. The purging volumes of inert gas are varied. Sorption is accomplished using a macroporous polymer (Spheron SE 50/50) followed by thermal desorption.[17] It is evident that under these conditions benzene can be completely extracted by bubbling approximately 2 liters of gas through the solution. However, the degree of extraction of *n*-heptylbenzene is approximately 60% and that of 2-methylnaphthalene is only 30%. For *n*-heptylbenzene the optimum volume of purging gas is twice as large, but the other components will be extracted only to the extent of 50%.

The selection of optimum conditions for the extraction from aqueous solution of polar organic compounds (alcohols, aldehydes, ketones, and

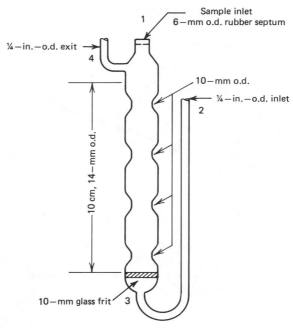

Figure 3.6. Device for stripping small volumes of water samples, according to Bellar and Lichtenberg.[19] *1*, Sample inlet (through a rubber septum); *2*, tube for the introduction of nitrogen; *3*, medium porosity glass frit (10 mm dia.); *4*, exit of vapors.

ethers with large distribution coefficients),[18] which are easily soluble in water, is an even more complex problem. Such details as the design and size of flasks to be used for purging the solution are of considerable concern. A special investigation into several types of instrumentation for stripping[18] indicated that the best extraction of water solutions is accomplished in a container recommended by Bellar and Lichtenberg[19] (Fig. 3.6). To extract the simplest carbonyl compounds, 2-butanol and 1-pentanol, from 8 ml of a 0.0007–0.08% solution with a yield greater than 90%, the system must be heated up to 95°C and purged with helium over a period of 30–60 min at a rate of 20 ml/min. Under such conditions methanol and ethanol are extracted with a yield of 30–50%. Increasing the container volume to strip samples of large volumes leads to a decrease in the effectiveness of the method. Therefore, to strip 100–200 ml of a water solution, it is more expedient to use a 1-liter flask with a magnetic stirring bar as shown in

3.1 ANALYSIS OF WATER AND AQUEOUS SOLUTIONS 111

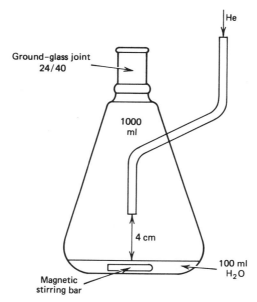

Figure 3.7. Device for stripping 100–200 ml water samples.

Figure 3.7.[18] When heated to 95°C, methanol and the simplest carbonyl compounds are extracted from a 0.0016% solution by 95%. However, acetic and butyric acids are not extracted.

The best adsorbent for the vapor concentration of organic compounds in the stripping of water solutions is Tenax-GC (poly[2,6-diphenyl-*p*-phenylene oxide]):

This heat-resistant hydrophobic organic polymer was originally cited for use in chromatographic columns.[20] Following the investigations of Zlatkis and co-workers,[21,22] the polymer is used for the concentration of organic atmospheric impurities and for the head-space analysis of biochemical samples when the separation of volatile organic components from a large quantity of water vapor is required. However, while hydrocarbons and certain oxygen compounds are totally adsorbed on Tenax, the simplest alcohols are retained poorly, but C_3–C_5 alcohols, aldehydes, and ethers are retained at 75–90%. The thermal desorption at 200°C is also usually accompanied by a 10–20% loss. This indicates that the total degree to which volatile organic substances can be extracted using Tenax and thermal desorption does not exceed 70–90%. It might be noted that the value for methyl, ethyl, and propyl alcohols is much lower. Thus, in quantitative head-space analysis of trace organic substances in water using extraction, the effectiveness of stripping, sorption and desorption must be determined experimentally. To this end, detailed methods have been developed. The first to be indicated refers to the previously mentioned work of Bellar and Lichtenberg.[19] The method described by them is applicable to the quantitative analysis of various water samples for the content of organic compounds with solubility less than 2% and a boiling point up to 200°C in concentrations from 10^{-7} to 25×10^{-4}%.

The processes including the vapor-phase concentration of impurities on the sorbent, its thermal desorption, and subsequent introduction into a chromatograph can be automated. Hewlett-Packard Coroporation has manufactured the HP-7675A purge-and-trap sampler for gas-chromatographic analysis of water. This system is to be used in determining the content of hydrocarbons and halogen derivatives. It automatically performs the stripping operations, the sorption on Tenax, the quick heating of the sorbent, the introduction of the vapor, and the preparation of the system for the next analysis.[23] The timing of the individual steps and the temperature of heating the sorbent and gas ports are programmed. Figures 3.8 and 3.9 illustrate the sensitivity of the analyses on such automated instrumentation. Figure 3.8 reproduces the chromatogram of a water sample (5 ml) containing 3×10^{-4}% gasoline. Clearly, about 70 hydrocarbon peaks are registered. Figure 3.9 is a chromatogram of tap water. The main peaks belong to chloroform and bromodichloromethane, the quantities of which totally satisfy official American standards.

In the final stage of vapor-phase concentration, extraction of the sorbate using volatile solvents can be employed instead of thermal desorption. This

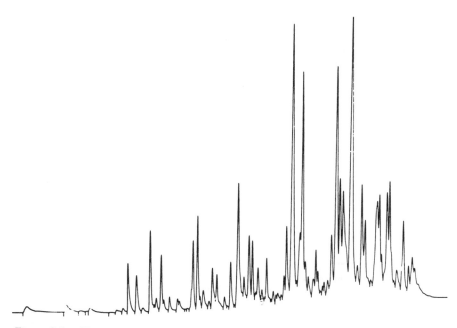

Figure 3.8. Chromatogram of a water sample contaminated with gasoline (0.015 mg in 5 ml), obtained using the HP-7675A accessory for automatic head-space analysis. Flame-ionization detector.

method is used in a technique developed by K. Grob and Zürcher.[24] The technique is used in many laboratories performing water analyses. Characteristically, this method completes the stripping in a closed (cyclic) system in which a limited volume of inert gas is recycled using a compression pump (Fig. 3.10). In this way the chance of contamination during the work is decreased. Recommended solvents for desorption include carbon disulfide

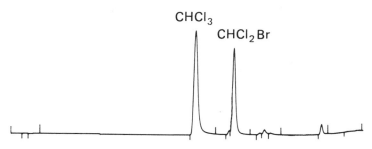

Figure 3.9. Chromatogram of trihalomethanes in tap water, obtained using the HP 7675A accessory for automatic head-space analysis. Electron-capture detector.

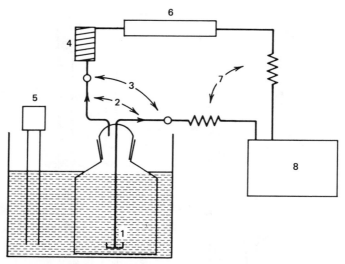

Figure 3.10. Schematic of a circulation apparatus for stripping large volumes of water. *1*, Glass filter (No. 0); *2*, Teflon tubes; *3*, glass-metal connections; *4*, heater; *5*, thermoregulator for the water bath; *6*, tube containing the adsorbent; *7*, stainless steel tubes (2 mm i.d.); *8*, pump.

and methylene chloride. The latter is less effective but is purified more easily. The limitation of the method is related to the use of the solvents. It is impossible to determine the most volatile components, as their chromatographic peaks are overlapped by the peaks of the solvents. As in other similar three-stage methods of extraction and concentration, all stages are closely interrelated and the conditions must be observed in all details. One-half to two liters of water at 30°C can be stripped by passing it at a rate of 1–2.5 liters/min over 2 hr.* Impurities are adsorbed on a disk of pure activated charcoal (1.5–5 mg) 2.7–3.6 mm in diameter and 0.7–1.5 mm thick (the dimensions depend on the quantity of impurities). The extraction using 1.5 mg of a sorbent is done with 5–15 μl of carbon disulfide or (in the case of more contaminated water) 10–100 μl of methylene chloride. The incompleteness of extraction is taken into consideration in the given method using internal standards. The 1-chloroalkanes C_6, C_{10}, C_{14}, C_{18} are used as the standards and are added to the sample as 5-μl aliquots of very dilute acetone

---

*If one is interested only in low-molecular-weight impurities (up to C_8), the time of stripping can be shortened to 15 min and that for collecting heavier substances can be increased accordingly.

solutions (1 : 100,000–1 : 20,000). The 1-chloroalkanes are convenient standards because of their availability. Rarely found in natural samples, they resemble many organic substances in their behavior during stripping and extraction. Direct comparison of the peak areas of the sample components with those of the standards gives a rough estimate of concentration. More correct results are obtained by calibration using water samples with exactly known quantities of the substances being determined and of the internal standards being added.

Stripping followed by cryogenic concentration, that is, head-space analysis with total recovery of the volatile components including the water vapors by low-temperature cooling, is rarely used, since the large quantities of water in the entrainment separator will disturb the subsequent (especially mass-spectrometric) investigation of the concentrate. The work of Novák and coworkers concerning the analysis of organic impurities in drinking water[25] can serve as an example of cryogenic concentration.

Determination of hydrocarbons in seawater using gas extraction can be completed with the stripping apparatus originally suggested by Wasik.[26] Actually, it is an electrolyzer for several liters of water (Fig. 3.11). The larger section of the electrolyzer, which is filled with water, has a spiral cathode constructed of gold wire separated from the anode by a macroporous glass filter of 30-mm diameter. To increase the sensitivity and accuracy of an analysis, the entire electrolyzer is placed into a thermostatted system and heated to $80 \pm 0.02°C$. The smallest hydrogen bubbles liberated from the cathode during the electrolysis rise through the volume of analyzed water. They reach equilibrium with the water and extract a portion of the volatile impurities. The use of hydrogen electrolyzed from the analyzed solution as the gas extractant totally excludes the possibility of contamination. It also allows the determination of negligible concentrations of gasoline hydrocarbons in concentrations as low as parts per trillion (10^{-12} ppt). For this, it is necessary to employ the concentration of the vapors and supply hydrogen from the upper section of the electrolyzer into the circulation system using an entrainment separator containing 2–3 mg of activated charcoal. Further analysis is completed according to Grob's method.[24]

During the analysis of gasoline in the order of 10^{-6} ppm, the hydrogen liberated by the electrolytic cell can be introduced directly into the chromatograph. This method employs the determination of the change in hydrocarbon concentration after purging with a certain volume of hydrogen as opposed to stripping gas extraction. In Wasik's paper[26] the basic equation of

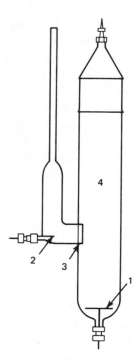

Figure 3.11. Electrolytic cell for the extraction of trace hydrocarbons from seawater using hydrogen. *1*, Cathode; *2*, anode; *3*, porous glass membrane; *4*, seawater sample.

continuous gas extraction computing the initial concentration of the impurities in a water sample is given, as is a formula to correct the volume of hydrogen collected for an analysis if the volume is sufficiently large. There are no data reported concerning the accuracy of such dynamic head-space analysis.

In general, the possibility of analyzing aqueous solutions using incomplete gas extraction is not well investigated. It has been recommended that when stripping gas extraction encounters the above-mentioned difficulties, dynamic head-space analysis should be used in the determination of the simplest oxygen-containing compounds.[9,27] One case was subjected to an experimental check based on extracting 5 ml of the investigated solution with wet nitrogen (3–30 liters) at 15°C. The equilibrium vapor (before and after the bubbling of nitrogen) was analyzed in the apparatus shown in Figure 3.1. The computation of the initial concentration of the analyzed solution was carried out according to Equation (1.20). The accuracy of the determination of ethyl, n-propyl, and n-butyl alcohols and of acetone and methyl butyl ketone in aqueous solutions in concentrations of 20–1300 mg/liter was less than or equal to 10% relative units.[9,27]

3.2 DETERMINATION OF VOLATILE ORGANIC SUBSTANCES IN BIOLOGICAL SYSTEMS

One of the most important applications of head-space analysis is the determination of ethanol in blood. The importance of this application lies in its use in toxicology and forensic chemistry. A great number of original articles and reviews[28–30] have been devoted to the substantiation, development, and perfection of the method. Until the 1960s, ethanol in blood was determined by the chemical analysis of venous blood, which does not satisfy modern medical and especially forensic chemistry requirements because of its poor selectivity[31] and very laborious and lengthy techniques.

The high sensitivity in the determination of ethanol in the blood of a person for medical purposes is dictated by the level of biological concentrations, which is in the order of several milligrams per liter.[32] In criminology the minimum concentrations determined are considerably higher, since ethanol present in blood in concentrations less than 100 mg/liter is considered within normal limits.[33] These concentrations can be measured by the direct chromatographic introduction of 1–3 μl of whole blood, plasma, or serum.[28–30] Due to its simplicity this method is fairly popular in clinical and forensic analysis. However, the direct introduction of biological fluids gives rise to fairly considerable experimental complications connected with contamination and the possible failure of the chromatographic column, overlapping peaks on the chromatogram, and defective syringes.[34] Jain[35] attempted to bypass these difficulties by inserting a small fiberglass plug into the head of the gas-chromatographic column. Nevertheless, even if this plug and several centimeters of the front of the column remain unchanged after 15–20 introductions, there are still foreign products accumulating that can lead to distortions in subsequent analyses.[36]

The generally accepted HSA method eliminates the shortcomings of the direct introduction of blood samples into a chromatographic column.[28–30] It must be noted that the principle of head-space analysis was used in conjunction with chemical analytical methods in the determination of alcohol in blood and urine by Harger et al. in 1950.[36a] Although the gas-chromatographic method they recommended is now outdated, the measured values of the distribution coefficients of ethanol between air and water, whole blood or urine still hold a certain interest. Their data showed that the distribution coefficients for whole blood are by 25%, and for urine by 8–9%, lower than for water. Data on the distribution coefficients of ethanol in various blood samples have been presented, including deviations from the standards; it appears that the variation in the values of K does not exceed 5–10%.

In the gas-chromatographic analysis of the alcohol content of blood, the static head-space method is used. Small glass bottles or vials closed with rubber caps are used as equilibration containers. Samples of equilibrium vapor are introduced into the chromatograph by means of gas syringes. Goldbaum and coworkers[37] recommended the combination of equilibration and introduction of the vapor into the chromatograph using medical syringes used to take human blood. The best reproducibility in sampling of the equilibrium vapor in the determination of blood alcohol content is accomplished in a specialized apparatus, such as the Alco-Analyzer (Luckey Laboratories, Inc.),[38] which uses heat conductivity at thermistors as the detector, and the universal head-space analyzers F-40, F-42, and F-45 manufactured by Perkin-Elmer. The pneumatic system of automated introduction of the equilibrium vapor into the chromatogram, as discussed in Chapter 2, was originally developed for these analyses[39] based on Machata's method.[40,41]

The methods described in the literature differ mainly in the conditions of conducting the preliminary operations, namely, in sample preparation. The main problems of head-space determination of ethanol in blood are connected with the increase in sensitivity of the analysis and the stabilization of the alcohol concentration in the samples selected. The volume of the blood sample has little influence on the sensitivity due to high values of K. Increasing the ratio of the phase volumes in an equilibration container up to V_G/V_L of 100 or even several hundreds insignificantly decreases the value of C_G [compare Equation (1.20)]. For example, increasing the ratio V_G/V_L from 8 to 350 leads to a decrease of only 13% in the height of the alcohol peak.[34]

Many methods assume the establishment of an equilibrium distribution of the alcohol between liquid and gas at room temperature (25–30°C). This

3.2 VOLATILE ORGANIC SUBSTANCES IN BIOLOGICAL SYSTEMS 119

only allows the determination of concentrations exceeding 100 mg/liter.[32] The detection limit can be significantly improved by placing the undiluted blood in the vials. This is related to the large values of ethanol distribution coefficients in aqueous solutions. To increase the sensitivity of the alcohol determination, salting out and heating the sample to 60–85°C during equilibration can be used to decrease the values of the distribution coefficients.[28]

The effectiveness of adding various salts varies. According to the data of Machata,[42] who investigated the salting-out ability of ammonium sulfate, sodium chloride, potassium carbonate, ammonium chloride, and sodium citrate at 60°C, the degree of increase in the concentration of ethanol in the equilibrium vapor fluctuates within the limits of 2–8 times, and the best results are obtained with potassium carbonate. According to Machata,[42] salting out must be done very carefully and only when necessary (when the ethanol concentration is less than 100 mg/liter). The addition of mineral substances can lead to the coagulation of blood and instability in the value of the distribution coefficient.

Using increased temperatures together with salting-out procedures allows the positive determination of ethanol in blood at biological concentration levels. However, the method can reduce the reproducibility and accuracy of the analysis[34,40] due to the rapid oxidation of ethanol to acetaldehyde, which is catalyzed by hemoglobin.[43,44] Reproducibility can be worsened by condensation and sorption on the walls and by loss of sample during the introduction of the equilibrium gas by common syringes into the evaporation chamber of the chromatograph.

The complications related to the use of increased temperatures were overcome by Wilkinson and coauthors,[32] who developed a method of determining ethanol at 60°C in the concentration range of 3 mg/liter to 1.2 g/liter that requires the analysis of only 20–50 μl of blood. The ethanol concentration in a selected sample is stabilized by the addition of sodium nitrate (to suppress the oxidation) and sodium fluoride (which prevents the accumulation of "false" ethanol that is produced by microorganisms during storage of the blood). Good analytical accuracy (±4.6%) is achieved by introducing the equilibrium gas into the chromatograph with a special 2 ml gas syringe thermostatted at 67°C.

With the addition of sodium nitrate, blood samples can be stored at –17°C over several weeks. The time for equilibration between blood and air at 60°C is approximately 3 min. The effect of blood group on the accuracy of analysis does not exceed the analytical error. According to Luckey's[38] data the relationship between the peak area and the ethanol concentration in the blood is linear up to 0.4%. The results of Glendening and Harvey[34] increased this limit to 4%.

The quantitative determination of ethanol in blood is done as a rule using an internal standard such as acetone, methyl ethyl ketone, or dioxane. Most frequently *n*-propyl alcohol (NPA) and *tert*-butyl alcohol (TBA)[28,32,42,45] are used, because they do not interfere with the gas chromatographic analysis: the ethanol peak is fully separated from their peak even in rapid analysis (Fig. 3.12). Also, NAP and TBA cannot be present in human blood. Isopropyl alcohol is not recommended as a standard for the analysis of capillary blood, as it is used to disinfect the surface of the skin before selecting the sample and may therefore be contained in the blood sample. In forensic chemistry, isopropanol is not used, since its "aging" during storage can lead to an increase in the ethanol content.[45]

The application of internal standards to biological samples requires the consideration of additional factors related to the characteristics of head-space analysis. Differing from the traditional method using the internal standard (in which a solution with a standard is introduced directly into the chromatograph), this technique requires consideration not only of the sensitivity of the chromatographic detector toward ethanol and the standard, but also of the type of medium being analyzed—whole blood, plasma, or serum[46] (for the difference in the values of the distribution coefficients see

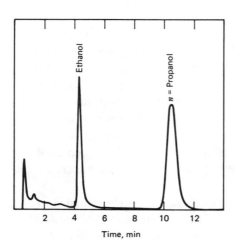

Figure 3.12. Typical chromatogram of the head-space of a blood sample analyzed for ethanol. The ethanol concentration of the sample was 0.6 mg/ml; the blood sample was mixed with equal volume of the internal standard solution containing *n*-propanol (1 mg/ml). Column: 183 cm × 3.5 mm i.d., containing Porapak Q 80–100 mesh. Column temperature: 150°C. Carrier gas (nitrogen) flow rate: 30 ml/min. Sample size: 0.3 ml. Flame-ionization detector; full-scale response: 1.5×10^{-10}A.

Section 1.4). The absolute value of the concentration of the standard can be absent in the calculations if the content of the standard remains constant and the conditions of sample preparation are strictly reproduced (with respect to the quantity of added reagents, temperature, and ratio of the phase volumes in the equilibrium container). The ethanol content in the blood sample is then determined by the previously completed calibration $A_e/A^{st} = f(C_e^0)$, where A_e and A^{st} are the peak areas of ethanol and the standard, respectively, and C_e^0 is the ethanol concentration in the solution.

Using an internal standard considerably decreases the requirements regarding the stability of temperature and concentration of salting-out reagents and eliminates the necessity of introducing equal volumes of the equilibrium gas into the chromatograph. Thus, according to the data of Hayck and Terfloth,[47] increasing the temperature of the equilibration container by 1°C increases the peak heights of ethanol and the standard by 5.5%, but the relationship of peak heights remains practically unchanged. Machata[42,46] showed that the vapor pressure curves (which depend on the concentration) for solutions of ethanol and *tert*-butanol above 0.1% in the temperature range of 45–70°C are parallel so that the relationship between the concentrations of these alcohols in a equilibrium gas for the indicated temperatures does not change (Fig. 3.13). A similar effect is observed during the change in the concentration of sodium fluoride.[47]

The chromatographic conditions must allow the complete separation of

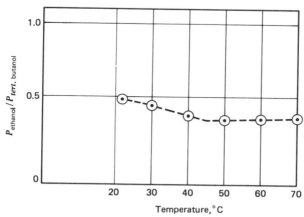

Figure 3.13. Relationship of the ratio of the vapor pressures of ethanol and *tert*.butanol in the headspace to the temperature of the sample vial. Sample: 0.1% solution.

ethanol and the accompanying volatile substances of blood in the shortest time of analysis (this is especially important in bulk determinations). Polyethylene glycol 1500 (PEG 1500), Halcomide 18, and Porapak Q[28,42] have acquired great popularity as stationary phases. The data in Table 3.2, in which the retention times of the toxicologically important substances are given, indicate that if only ethanol is present in the blood the minimum time of analysis (approximately 3 min) using *tert*-butanol as the standard is achieved with PEG 1500. To attain this, the efficiency of the column must correspond to at least 800 theoretical plates. Halcomide 18 and Porapak Q increase the length of the analysis to some degree, but it still does not exceed 10 min.

The computational parameter of the chromatogram is the peak height or area (using an integrator). The relationship between the peak heights and areas of the analyzed substances and the standard can considerably differ,[42] apparently, due to the asymmetry of the peaks. Each of these methods allows

Table 3.2 Retention Times (Minutes) of Certain Toxicologically Important Substances[a,b]

Substance	Retention Time, min		
	Polyethylene glycol 1500(90°C)[b]	Halcomide 18 (90°C)[b]	Porapak Q (175°C)[b]
Methanol	2.1	2.6	1.1
Ethanol	2.6	4.0	2.1
Isopropanol	2.5	4.8	3.4
n-Propanol	4.2	8.7	4.5
n-Butanol	7.2	19.3	10.2
Acetaldehyde	1.1	1.7	1.5
Diethyl ether	0.8	1.3	4.2
Acetone	1.5	2.1	3.2
Chloroform	2.8	2.4	7.2
Formaldehyde	4.6	12.7	4.3
Methyl ethyl ketone	2.1	4.2	6.9
tert-Butanol	2.2	5.4	5.4

[a]The length of the column is 2 m, the column packing consists of 15% of the stationary phase on Celite, and the carrier gas (nitrogen) flow rate is 24 ml/min.

[b]Column temperature.

3.2 VOLATILE ORGANIC SUBSTANCES IN BIOLOGICAL SYSTEMS

accurate measurements in the analysis,[48] but for determining concentrations near the threshold sensitivity of the method (increased noise levels of the chromatographic detector), peak heights are the preferred measurement. When a stable zero signal is evident, it is more convenient to use an integrator, as the total automation of the computation becomes possible.

High accuracy of the alcohol determination in blood or serum (2–3%) is acquired using the Multicraft F-45 automatic head-space analyzer, due to its excellent sampling reproducibility (up to 1%) of the equilibrium gas into the chromatograph.[39,49] Calibration is done using standard aqueous solutions with a known amount of ethanol and a constant (but not computationally included) concentration of the internal standard (*tert*-butanol).[42,48,49]

The following outlines the procedure of the analysis. The equilibrium distribution is established at 60°C in small glass vials of 24-ml volume. The blood samples (0.5 ml) to be analyzed are introduced into the vials using a pipet. One of the containers is used for the calibration of the apparatus. To this end, 0.5 ml of a standard ethanol solution (0.1 or 0.2%) is placed in this container. Then, 0.1 ml of the *tert*-butanol solution (0.2%) is added to all samples, including the standard. All vials are hermetically sealed using silicone rubber septa. After establishing equilibrium, approximately 0.5 ml from each vial is automatically introduced into the chromatograph. Figure 3.14 illustrates chromatograms obtained from a serial determination.

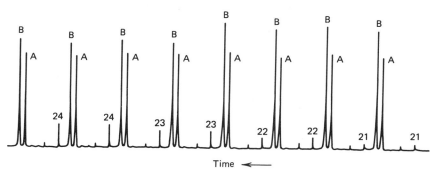

Figure 3.14. Chromatograms obtained in analyzing a series of blood samples with the Perkin-Elmer F-42 head-space analyzer. The concentration of *tert*.butanol (internal standard) was identical in all the samples. *A tert*. butanol peak; *B* ethanol peak. The number at the start of each chromatogram indicates the sequential number of the sample vial. Column: 2 m × 0.25 in. o.d., containing 15% Carbowax 1500 on Celite. Column temperature: 70 ml/min. Flame-ionization detector.

From the results of the analysis of the standard solution, the calibration factor is determined as

$$f_k = C_e^{st} \frac{A_b^{st}}{A_e^{st}} \qquad (3.2)$$

where C_e^{st} is the concentration of ethanol in the standard solution, and A_b^{st} and A_e^{st} are the peak heights or areas of *tert*-butanol and ethanol. The calibration factor f_k takes into consideration the experimental conditions of the analysis, the relationship of the phase volumes in the equilibration container, and the value of the distribution coefficients of ethanol and the standard. In addition, the numerical value of f_k is determined not only by the concentration of ethanol in a standard solution, but also by that of *tert*-butanol. Therefore, the preparation of samples for analysis demands a constant concentration of the standard substance in both the standard solution and the investigated samples.

The alcohol concentration of the investigated blood sample (C_e^0) is calculated using the equation

$$C_e^0 = \frac{f_k}{F_k} \frac{A_e}{A_b} \qquad (3.3)$$

where A_b and A_e are peak areas or heights of *tert*-butanol and ethanol, and F_k is the factor that accounts for the difference in the distribution coefficients of the determined substances in an investigated solution in comparison with the aqueous solution. Machata[42,46] determined the numerical value of F_k by measuring the relationship of the peak heights of a standard in the investigated samples and in water at equal concentrations in both solutions. However, in accordance with the above-described procedure of analysis, the value of F_k must also consider the change in the distribution coefficient of ethanol. Proceeding from the basic equation of head-space analysis using an internal standard (1.48), it is easily shown that the value of F_k is determined by the expression*

$$F_k = \frac{A_b^{st}/A_b}{A_e^{st}/A_e} \qquad (3.4)$$

---

*In the equilibrium container it is assumed that the ratio of the phase volumes is much less than the ratio of the values of the distribution coefficients of ethanol and *tert*-butanol at equal concentrations in solutions.

3.2 VOLATILE ORGANIC SUBSTANCES IN BIOLOGICAL SYSTEMS

Therefore, the values of $1/F_k$ given by Machata[42,46] and Battista[50] (for whole blood: 1.08 and for serum: 1.26), are approximate, but the degree of approximation is apparently sufficient for achieving the required analytical accuracy.

When analyzing a large number of whole blood and serum samples the calibration factors can be computed from the results obtained from the standard samples placed periodically between the actual blood samples and the peak area (heights) of the internal standard in each sample. The calibration factors depend linearly on the relationship of the peak heights (areas) of the standard in the samples and in water (at equal concentrations).[50]

The volume of the blood sample required in automatic head-space analyzers can be decreased to 0.1 ml.[51] Simultaneously, the volume and concentration of the added standard (the solution of *tert*-butanol) must increase to 0.5 ml and 0.5%. The factor $1/F_k$ under such conditions is 1.85, 1.71, and 1.59, for whole blood, plasma, and serum, respectively.

Due to its simplicity, high accuracy, and reliability, and the possibility of shorter times for completing large numbers of analyses using total automation for computation, the procedure using the automatic Perkin-Elmer head-space analyzers for determining ethanol in blood is officially accepted in the majority of European countries.[48]

The concentration of alcohol in blood usually serves as a measure of its content in the organism.* The method of collecting blood from a person is an important consideration as the content of the ethanol in the arteries is 1.5–2 times higher than in the veins. It has been recommended[28,32] that capillary blood from a finger be collected since it is similar to arterial blood. The analysis of capillary blood, when necessary, allows the collection of larger numbers of samples and the use of less qualified personnel. As an example, the possibilities of head-space analysis can be illustrated by the dynamics of the change in the ethanol concentration in the blood of an adult male (Fig. 3.15).[32] The measurements were taken each 10–15 min over a 3-hr period after ingestion of 15 ml of 95% ethanol mixed with 150 ml of orange

---

*The ethanol content in blood is not a direct reflection of its concentration in the entire organism, because the alcohol is unevenly distributed in the various organs. The saturation of one or another tissue by alcohol increases with the increase of the tissue's water content and decreases with an increase in the fat content. It is important that enzymes be present in the tissue to break down the alcohol.[33]

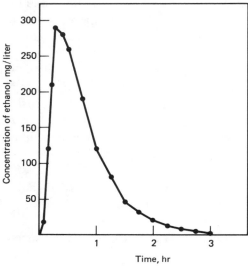

Figure 3.15. Change of the concentration of ethanol in capillary blood of an adult male after consumption of 15 ml of 95% ethanol mixed with 150 ml of orange juice, against time.

juice. The maximum concentration of ethanol in capillary blood (290 mg/liter) is achieved after 16 min, and after 3 hr the alcohol content drops to the biological level (2.4 mg/liter).

The HSA methods of determining ethanol in blood discussed above can also be applied to urine, without any major change. The ethanol content of urine is widely used in forensic-medical applications, because the relationship of the alcohol concentrations in blood and urine helps to establish the stage of intoxication and thus to evaluate the time and extent of alcohol consumption. Besides that, in the doping test of athletes urine samples are analyzed. For certain sports the ethanol content is tested.

The direct head-space analysis of blood and urine also allows the determination of the content of other aliphatic alcohols, which leads to the establishment of the degree and type of alcoholic intoxication. The relationship of the concentrations of *sec*-butyl and isobutyl alcohols to *n*-propanol permits certain conclusions concerning the nature of the alcoholic beverage consumed (beer, wine, brandy, or wheat vodka).[52] Figure 3.16 illustrates an example of head-space analysis of aliphatic alcohols .[53]

Equilibrium vapor analysis is used not only for the determination of alcohols but also of other toxicologically important substances in the

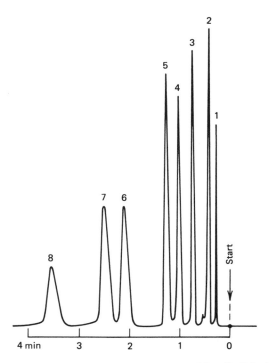

Figure 3.16. Chromatogram illustrating the rapid separation of $C_1 - C_4$ aliphatic alcohols with help of the Perkin-Elmer HS-6 accessory for head-space analysis to the Sigma gas chromatographs. Column: 100 cm × 3 mm i.d., containing 0.4% Carbowax 1500 on graphite 60–80 mesh. Column temperature: 100°C. Flame-ionization detector. *1*, Methanol; *2*, ethanol; *3*, propanol-2; *4*, propanol-1; *5*, 2-methylpropanol-2; *6*, butanol-2; *7*, 2-methylpropanol-1; *8*, butanol-1.

biological context.[54,55] Those substances include acetone, acetaldehyde, anesthetics (ether, chloroform, halothane), the bases of amphetamines, halocarbons[56–58] and aromatic hydrocarbons,[59] methylmercaptan,[60] and methyl methacrylate.[61] In most cases, for volatile substances in biological liquids, the technique and methods for quantitative analysis are similar to those discussed for ethanol. Basically, the differences concern the conditions of the gas-chromatographic separation, the selection of a standard, the equilibration temperature, and the methods of introducing the gas phase into the chromatograph.

The analysis of solid objects (pieces of tissue, organs) requires additional operations in sample preparation. For this purpose extraction by a solvent or steam distillation is used. The solution obtained in the organic solvent or water is analyzed in the same manner as a biological fluid. However, when evaluating the accuracy of the analysis, it is necessary to consider the completeness of substance extraction from the solid tissue. A typical case is the determination of acetone,[62] aromatic hydrocarbons,[63] and chlorinated aliphatic hydrocarbons[64] in biological tissues. The components are extracted by steam distillation, and the gas phase over the condensate is then introduced into the chromatograph. Sometimes the content of the volatile substances in solid objects is determined by analyzing the gas over the investigated sample, which exists as a suspension in a solvent. An example of this application is illustrated by official Soviet forensic methods of the quantitative determination of dichloroethane in a corpse.[65] An accurately weighed 5-g mass of the investigated object (an average sample is taken from 100 g of well-ground matter) is placed into a vial (penicillin type) to which 3.5 ml of 96% ethanol and 0.5 ml of a standard alcoholic solution of benzene (1.7 mg/liter) are added. The solution is mixed well, the vial is sealed with a rubber septum and placed into a metal cylinder, two-third of which is immersed in a boiling water bath for 5 min. With a syringe heated to 60°C, 5 ml of the gas phase is collected and introduced into the chromatograph. The peak heights in the chromatogram of the substance being determined can be related to the standard through a previously constructed calibration curve (under conditions identical to the analysis). This allows the calculation of the concentration of dichloroethane in the investigated sample. The conditions of the gas-chromatographic analysis are: column 240 × 0.6 cm packed with 15% triethylene glycol on spherochrome-1 (0.2–0.3 mm), temperature 96°C, carrier gas (helium, hydrogen) flow rate 70 ml/min; a catharometer is used as the detector. The detection limit is 0.25 mg in 5 g of sample. The determination error in the dichloroethane concentration range of 0.25–12.5 mg is 5–10%. The same method was developed by Kojima and Kobajashi[59] while determining the toluene content in tissues over the concentration range of 0.2–2 mg/liter. The only difference was that the solid tissue was suspended in water, and ethylbenzene in ethanolic solution was added to the suspension as the standard.

The determination of the volatile substances in biological liquids is sometimes done by the practically complete extraction of the analyzed components into the gas phase. Natelson and Stellate[66] developed a method and the necessary instrumentation for the determination of acetone, methanol, ethanol, and isopropanol in urine. The principal difference of this method from the one just described consists of the following. Before gas-

3.2 VOLATILE ORGANIC SUBSTANCES IN BIOLOGICAL SYSTEMS

chromatographic analysis, the investigated sample (0.1 ml) is almost completely dehydrated at 50°C by introducing 0.25 g of calcined sodium sulfate, copper sulfate, or cadmium sulfate. As a result, the substances being determined are almost totally transferred into the gas phase. The reaction container is then subjected to the flow of the carrier gas and the volatile components are introduced into the chromatographic column. The total extraction of the volatile substances from a biological sample (not less than 94%) and their single introduction into the chromatograph increase the sensitivity of the analysis and allow the application of the catharometer as the detector, for concentrations of 10–30 mg/liter and higher. This considerably simplifies the instrumentation of the method, which is important when organizing the analyses for the conditions of clinical laboratories.

The principle of the complete extraction of the converted products of the analyzed substance from a liquid sample (blood, serum, urine) into the gas phase was instituted as the basis of a routine method for the determination of aliphatic C_1–C_5 alcohols, as suggested by Pomerantsev,[67] and for carboxyhemoglobin, as developed by Blackmore.[68]

The alcohol determination accepted as forensic practice in the USSR is based on the quantitative transformation of the alcohol into the alkyl nitrites, which are volatile and poorly soluble in water. The reaction is carried out in glass vials (penicillin type), hermetically sealed, using 0.25 ml of a 30% aqueous solution of sodium nitrite introduced into the mixture of 0.5 ml of a 50% solution of trichloroacetic acid and 0.5 ml of the investigated solution. Within 1–2 min after mixing the reagents, the alkyl nitrites are almost totally extracted from the concentrated sodium trichloroacetate solution, and the gas phase from this container is introduced into the chromatograph. The time of the chromatographic determination is 6 min (Fig. 3.17). The error in the determination using an internal standard (an alcohol) is ±3–5%.

The content of carboxyhemoglobin in whole blood and in tissue extracts is determined by the quantity of carbon monoxide liberated during its decomposition by potassium ferrocyanide at pH 8. The concentration of carbon monoxide in the gas phase is measured using a molecular sieve 5A (50–80 mesh) column and a catharometer detector.

Volatile product analysis of blood and urine is also used to determine the content of chloralhydrate, paraldehyde, trichloroacetic acid, and trichloroethanol. The latter two substances form in the body upon poisoning by trichloroethylene. The determination of chloralhydrate[37] is based upon

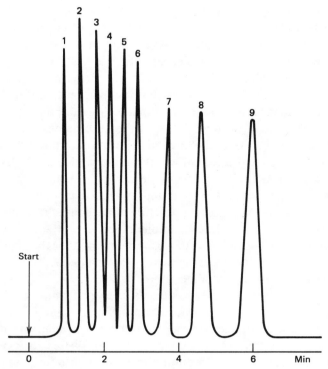

Figure 3.17. Chromatogram of alkyl nitrites. Column: 200 cm × 6 mm i.d., containing 25% triethylene glycol on INZ-600 (0.20–0.25 mm). Column temperature: 75°C. Carrier gas (nitrogen) flow rate: 35 ml/min. Thermal-conductivity detector. *1*, Nitrogen oxides; *2*, methyl nitrite; *3*, ethyl nitrite; *4*, isopropyl nitrite; *5*, n-propyl nitrite; *6*, isobutyl nitrite; *7*, n-butyl nitrite; *8*, isoamyl nitrite; *9*, n-amyl nitrite.

its alkaline hydrolysis and quantitative formation of chloroform. Paraldehyde in acidic medium is hydrolyzed (after adding one drop of concentrated sulfuric acid) to acetaldehyde. Trichloroacetic acid is determined as methyl ether[69] after the investigated sample is treated with dimethyl sulfate in a very acidic medium. For the determination of trichloroethanol, Lindner and Weichardt[70] used the conversion of the alcohol into chloroform (haloform sample) immediately before analysis. However, Triebig[69] achieved good results by direct head-space analysis of trichloroethanol with a determination limit of 0.1 mg/liter and an error of 5–12%.

3.3 DETERMINATION OF VOLATILE SUBSTANCES IN POLYMERS

The residual monomers and low-molecular-weight nonpolymerizing impurities that enter the polymer systems from the initial raw materials and from the solvents used in the production of polymers have an extremely adverse influence upon the characteristics of the polymers. Residual organic solvents in polymer films can also originate in the paint and varnish materials that are used for printed designs and inscriptions. Sometimes the volatile impurities can get into the plastics with the plasticizers. Finally, in some polymers used in the medical field, ethylene oxide, which is used in their sterilization, can be present in residual quantities. The majority of the volatile impurities contained in polymeric materials are dangerous and poisonous substances. Vinyl chloride monomer is a carcinogen, and inhalation can lead to cancer of the liver. The content of these components in manufactured items is strictly controlled. Especially strict standards are established for materials used in packaging and storing food products. Even volatile impurities with a comparatively low toxicity can get into food and considerably change its smell and taste, lower its quality, and make it useless.

The determination of trace volatile impurities has become one of the most important directions in the analytical chemistry of polymers. An application of head-space analysis to this purpose is very timely, because the introduction of polymers into a gas chromatograph is undesirable and is not always possible. However, head-space analysis of polymers requires the consideration of specific properties of the analyzed substances. The great majority are solid materials that are difficult to dissolve in common solvents and decompose at comparatively low temperatures. The simplest solution to the problem seems to be the analysis of the equilibrium gas phase over a polymer. However, the diffusion of the volatile components from a solid polymer to the surface is difficult and equilibrium is established very slowly. For this reason the first applications of head-space analysis to the analysis of polymers were based on solubility. Unfortunately, appropriate solvents do not exist for some commercially important polymers. Therefore, direct head-space analysis of solid polymeric materials remains an urgent problem. Also important is the problem of the analysis of polymer suspensions, which are widely used but cause additional complications characteristic of colloidal systems. Therefore, three cases of possible HSA application to the determination of volatile impurities of polymers should be discussed: the analysis of homogeneous solutions of polymers, of polymer emulsions, and of solid polymers.

Solutions of polymers. These can be subjected to direct gas chromatographic analysis to determine the volatile components by introducing them into the chromatograph directly or after reprecipitation with methanol. Such methods, which have been used for a long time, are officially accepted in many countries.[71-73] The main shortcoming lies in the necessity of frequently changing the chromatographic columns and cleaning the vaporization chambers contaminated by polymers. The direct introduction of the solution is sometimes impossible due to the overlapping of the broad solvent peaks with the peaks corresponding to the impurities. The introduction of polymer solutions becomes difficult due to their high viscosity and adhesion. In the vapor phase these difficulties disappear, and the relationship between solvent peaks and volatile impurities is much more favorable, especially if the solvent has a low vapor pressure.

The deciding criterion for selecting a solvent is its ability to dissolve the polymer. In addition to this, high-boiling solvents are preferred as they are easily purified and have longer retention times than the impurities. The solvents most often used are acetamide and dimethylformamide (see Table 3.3). The limit of sensitivity in such determinations depends heavily upon the volatility of the impurities. For gaseous monomers (vinyl chloride, butadiene) in the tabulated organic solvents, it reaches 0.05 ppm (instead of 1–5 ppm with the direct introduction of the polymer solutions). However, the sensitivity of head-space analysis of liquid monomers is considerably lowered, decreasing rapidly with decreasing volatility (increasing the boiling point). The impurity determination threshold of styrene (b.p. 145°C) using vapor-phase analysis of polystyrene solutions in dimethylformamide or dimethylacetamide is 10–20 ppm, that is, no better than the direct analysis of these solutions. In the case of the higher boiling 2-ethylhexyl acrylate (b.p. 214°C), the sensitivity of the head-space analysis is considerably worse (only 1000 ppm). Steichen[78] suggested the addition of 40–70% water to the solutions of the polymers, lowering the distribution coefficients of volatile impurities and increasing their concentration in the gas phase. With the introduction of water to solutions of styrene polymers and copolymers, the determination of residual styrene showed a 20-fold increase in sensitivity making it possible to determine styrene down to 1 ppm. In the case of 2-ethylhexyl acrylate (EHA), the addition of water improved the sensitivity

Table 3.3 Examples of HSA Solution Methods for the Determination of Volatile Impurities in Polymers

Polymer	Impurities Determined	Solvent	Ref.
Polystryene	Styrene, ethylbenzene, xylols, n-propylbenzene, cumol	Dimethylformamide	74
Polystyrene	Styrene	Dimethylacetamide	75
Poly(vinyl chloride)	Vinyl chloride	Dimethylacetamide	76–78
Styrene-butadiene copolymer	Butadiene	Dimethylacetamide	78
Styrene-acrylonitrile copolymer	Acrylonitrile	o-Dichlorobenzene	78
Polystyrene	Styrene	Dimethylacetamide + water (4:3)	78
Styrene–2-ethylhexyl acrylate copolymer	2-Ethylhexyl acrylate	Dimethylacetamide + water (2:5)	78
Poly(vinyl chloride)	1,1,1-Trichloroethane	Dimethylformamide + water	79
Acrylonitrile-butadiene-styrene copolymers	Acrylonitrile	Dimethylsulfoxide Dimethylformamide	80
Poly(methyl methacrylate)	Ethylene oxide	Acetone	81

200-fold, permitting the determination of 5 ppm EHA. (The temperature of the solution was 90°C in both cases).[78]

Generally, head-space analysis of polymer solutions has been completed using the automatic Perkin-Elmer analyzers and employing the methods of absolute calibration or internal standard. The relative error was 2–4%.[79] Substances chemically similar to the analyzed impurities (n-butylbenzene, ether, etc.) were used as the internal standard. Solutions used for absolute calibration should contain the corresponding polymer, since the presence of the polymer in solution at working concentrations (in the order of 10%) considerably influences the value of the distribution coefficient for the impurities. When standard solutions are being prepared and no source of pure polymer is available, it is recommended that a known quantity of the impurity being determined be added to the solution of the analyzed polymer.[82]

Polymers emulsions. Large quantities (millions of tons) of polymers are presently being manufactured and used as aqueous dispersions in paint emulsions and synthetic latexes, which are used in such rubber products as trimming, binding (adhesive, glue, cement), and building materials. The polymers usually comprise 40–60% of the colloidal systems existing as globules with an average diameter varying from 0.08 to 0.3 μm. For such materials, an important requirement is a low residual monomer content, as the monomers impart an undesirable smell and are toxic. As a rule, the allowable monomer concentration in synthetic latexes is 0.02–0.1%. In some cases it can be in the order of 0.5% (poly(vinyl acetate) and polyacrylate dispersions). The use of head-space analysis for the control of monomer impurities in aqueous dispersions of polymers seems very promising; however, direct head-space analysis of similar dispersions is a more complex problem than the analysis of homogeneous systems. First, the volatile components of latexes will distribute between three (not two) phases (the condensed dispersion phases and the vapor phase). This necessitates the consideration of two different distribution coefficients and the relationship concerning the quantities of the condensed phases. Second, the small size of the globules in the organic phase requires the investigation of the influence of such factors as the curvature and size of the dividing surface between the phases upon the results of the analytical determinations. These problems have not been the subject of detailed study.

Determining the apparent (average) distribution coefficient of the monomer between a dispersion and its vapors as a function of the total monomer concentration in a dispersion C_D and in the gas phase C_G will give

$$K_D = C_D/C_G \qquad (3.5)$$

It is not difficult to show that

$$K_D = K_1\psi_1 + K_2\psi_2 \qquad (3.6)$$

where K_1 and K_2 are the distribution coefficients between the dispersion phases and the gas phase, and ψ_1 and ψ_2 are the phase fractions ($\psi_1 + \psi_2 = 1$) by weight or volume, depending on the method of expressing the concentration C_D.

If the degree of dispersity is constant, then the influence of the curvature of the surface in the dispersed phase will be reflected in the numerical value of the corresponding distribution coefficient. From Equation (3.6) it follows that the coefficient K_D can retain uniformity either at $K_1 = K_2$, neglecting the dependence of the phase composition, or at fixed phase composition. In these cases the same simple proportional dependence between the concentrations of the volatile substance in the vapor and the dispersion itself is observed as in the systems with a homogeneous liquid phase. In some dispersions a linear dependence was achieved regarding the peak area of the monomer in the chromatograms of the vapor phase determining the concentration of its emulsion. For acrylonitrile, the proportionality of its concentrations in the vapors of the dispersion of its copolymers is retained down to a liquid content in the order of 1 g/liter.[82] The composition of the vapors of polyacrylonitrile dispersions proved sufficiently complex that the acrylonitrile had to be registered using a nitrogen-selective detector (Fig. 3.18).

To eliminate the complications related to direct head-space analysis of the polymer dispersions, Hachenberg[83] suggested the preliminary dissolution of the aqueous dispersions of styrene-acrylate base in dimethylformamide. The gas phase above the transparent solutions was then subjected to analysis using the Multifract F-40 with an internal standard (*n*-butylbenzene). The impurities butyl acrylate, butyl propionate, and styrene were determined. Due to the large quantities of water it was necessary to regulate the ratio of dispersion quantities to solvent (0.2 g and 3 ml) since the water that gets into dimethylformamide from the dispersion influences the distribution coefficients of the monomers between the solution and the vapor.

136 QUANTITATIVE DETERMINATION OF IMPURITIES

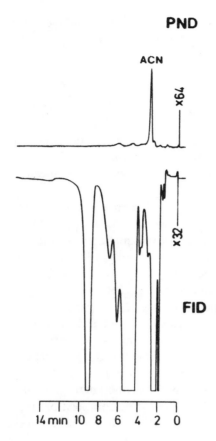

Figure 3.18. Head-space chromatogram from 1 ml of an aqueous suspension of acrylonitrile copolymer containing 2 mg/liter residual monomeric acrylonitrile (ACN). Perkin-Elmer F-42 head-space analyzer. Column: 2 m × 0.25 mm o.d., glass, containing 10% diethylene glycol sebacate on Chromosorb W AW. Column temperature: 70°C. Sample temperature: 80°C. Detector: top chromatogram, nitrogen-phosphorus selective detector (PND); lower chromatogram, flame-ionization detector (FID).

Adding water, as mentioned earlier, was used to increase the sensitivity of the determination of certain residual monomers in dimethylacetamide solutions of the polymers.[78]

Solid polymers. With large granules or flakes, the time needed for the establishment of gas-solid equilibrium is too long even at high temperatures (more than several hours). Therefore, head-space analysis of solid polymers in equilibrium or near-equilibrium conditions can be conducted only for finely dispersed substances and sufficiently thin films. Such limitations cannot be applied to the many objects being analyzed. Also, the wide distribution of polymer and plastic products for consumer, construction, and medical use requires analysis and control regarding the contamination of the environment by dangerous impurities contained in these products.

3.3 DETERMINATION OF VOLATILE SUBSTANCES IN POLYMERS 137

The specifics of such investigations require a change in the approach and in the goals of head-space analysis. The impossibility of achieving equilibration allows the use of other (kinetic) principles of analysis. The most important problem becomes not only establishing the concentration of the impurities in the substance, but also their accumulation in the contacting environment (in the air, water, food products, or tissues of living organisms).

Until recently, the detailed study of the conditions of the direct equilibrium vapor-phase analysis of solid polymers was carried out only for poly(vinyl chloride) in the works of Berens.[84-86] The time required for diffusion of a certain fraction of monomer from the spherical granules of the polymer into the vapor phase is proportional to the square of the granule diameter (d) and inversely proportional to the diffusion coefficient. (D) [85]:

$$t_{99} = \frac{0.113 d^2}{D} \qquad (3.7)$$

t_{99} is the diffusion time for 99% of the monomer contained in the granules.

For vinyl chloride in poly(vinyl chloride) powder, the diffusion coefficient at 90°C is $D = 2 \times 10^{-10}$ cm^2/sec.[85] Thus, the time required according to Equation (3.7) for evaporation of almost all of the monomer impurity from the granules of diameter less than 25 μm is approximately 1 hr. In approximately the same time, the PVC powder heated to 90°C will establish equilibrium with the gas phase. The solubility of vinyl chloride in poly(vinyl chloride) at temperatures higher than the glass transition point up to the concentration of 0.4% is proportional to its partial pressure in the vapor phase.[84] In this case the simplest formulas of HSA (as discussed in Chapter 1) are applicable. The content of residual vinyl chloride can be determined by introducing the vapor phase of weighed portions of PVC equilibrated in closed containers at 90°C over 1 hr into the chromatograph. When using the F-40 automatic analyzer for samples with vinyl chloride content of 0.5–750 ppm, the standard deviation in parallel determinations was 4.6%, several times less than during the analysis of PVC solutions. However, in the case of pressed, molded, or fused samples, the time of equilibrium is long and head-space analysis is unsuitable for the determination.[86]

In direct head-space analysis of solid polymers, calibration using standard samples of known composition is rarely applied, since the preparation of such solid samples with different and exactly known content of volatile impurities is very difficult or impossible. Approximate evaluations are most frequently used by employing conditions favorable for the diffusion of the largest part of the volatile impurities from a sample. This is accomplished by increasing the temperature and volume of the gas phase and disregarding the remaining fraction of the impurities in the polymer. Such an approach is justified in the area where head-space analysis of solid samples has acquired greatest popularity: in the determination of residual solvents and monomers in the polymer films used for the packing of food products. Optimum conditions of analysis are found empirically, and in the simplest HSA cases, samples from the gas phase above the solution are collected using a syringe without strict thermal control and without taking pressure fluctuations into account but retaining the same working mode while constructing the calibration diagrams. The determination of trace hydrocarbons in one of the popular packing materials can serve as an example of the method. Medium pressure polyethylene (MPE) is obtained by suspension polymerization in a hydrocarbon medium.[87] The granules or articles are placed in a glass container sealed with a silicone rubber septum and heated for 40 min at 120°C. With a syringe, 1 ml of the head-space gas is obtained for introduction into the chromatograph. The quantity of solvent released from the MPE is calculated from the calibration diagram constructed using several hydrocarbon samples (2–4 μl) equilibrated in the same containers, which are heated under identical conditions.

Similarly, one can control the content of residual methylene chloride in the packing material Escaplen.*[88] Pieces of the film, 4 × 4 cm square and 40 μm, thick are put into penicillin vials closed with rubber septum and are thermostated at 100°C for 10 min. The syringe is then flushed twice with the vapor–air mixture, and 2 ml of this mixture is introduced into the chromatograph. The sensitivity of the analysis is approximately 10 ppm.

Considerably higher sensitivity (to 0.05 ppm) was obtained in the head-space determination of trace ethylene oxide in sterilized medical materials.[89] Solid samples of 20–200 mg were equilibrated in 9-ml glass containers at 100°C. Heating these samples for 15 min was sufficient to convert almost all of the residual ethylene oxide into the gas phase. Samples of 0.1 ml were used for the gas chromatographic analysis. In comparison with extraction by solvents, the HSA method is more convenient, faster, more accurate, and easily automated. When a means exists for automatic sample collection, it is possible to do 50 determinations during a work day. The calibration is done

---

*Synthetic isoprene raw rubber with 10% plasticizer (dioctyl sebacate) and 0.1% sorbic acid is used as a stabilizer.

using gas mixtures with a known content of ethylene oxide. The analysis of the volatile impurities of solid polymers by means of evaporation into the gas phase using an absolute calibration with the same apparatus, but without weighed samples of the polymer, has been recommended by ASTM.[90]

Instead of the "absolute calibration," a unique type of internal standard analysis was suggested as there is a disregarded quantity of volatile component remaining after the head-space analysis of the polymer. A volatile standard is introduced together with the polymer into a thermostated and hermetically sealed container.[91,92] This technique apparently gives more reliable and more reproducible results when the distribution coefficients of the standard substance and the impurities are similar. It has been used for the determination of such solvents as ethanol, ethyl acetate, methyl ethyl ketone, 2-ethoxyethanol, propanol, and toluene in laminated polyethylene and polypropylene films coated with poly(vinylidene chloride).[91]

However, with larger samples of the solid polymer, the fraction of the volatile impurities changing into the gas phase rapidly decreases. Thus, in the head-space analysis of residual ethylene oxide in sterilized rubber gloves, only half of the ethylene oxide in large weighed samples (0.6 g) of the gloves changed into the gas phase even at 120°.[93] In some cases the degree of transition of the volatile impurities into the gas phase can be controlled through repeated extraction using fresh volumes of the gas, and the total content of the impurities in a polymer can be evaluated. This method was first suggested in 1970 for the determination of organic solvents in adhesive tapes by Suzuki, Tsuge, and Takeuchi,[94] and then repeatedly described in somewhat modified form by Kolb.[82,93] The Japanese researchers established that upon periodic substitution of the gas phase above the heated sample containing the volatile components of a polymer, the logarithms of the quantity of volatile components changing into the gas phase depend linearly upon the number of substitutions of the gas phase (Fig. 3.19). The slope of the plots increases when increasing the temperature of the polymer sample and decreases with increases in the boiling point of the volatile impurities. However, it also depends upon the specific interaction of the impurities and the polymer. The existence of a linear dependence in semilogarithmic coordinates makes it possible to evaluate the total content of the volatile impurities by summing up the peak areas for a considerably large number of extractions. These number of extractions are not necessarily determined experimentally because the total peak area can also be determined by

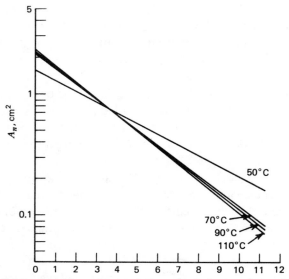

Figure 3.19. Relationship of the peak areas of n-hexane in the head-space above an adhesive tape to the number of consecutive substitutions (extractions), at various sample temperatures.

graphical linear extrapolation. The authors worked under conditions near equilibrium, and they were experimentally establishing the polymer heating time necessary for reacting a practically stable concentration of the volatile components in the gas phase. (For light hydrocarbon solvents, C_5–C_7, the time required was 16–18 min at 70°C.) However, in their assumed method of "periodic introduction," the gas phase is not necessarily a equilibrium. In particular, the straight line in Figure 3.19 corresponding to the extraction of hexane at 50° was obtained under nonequilibrium conditions.

The same principle of direct vapor-phase analysis of polymers is discussed in a subsequent article of Kolb and Pospisil,[93] whose handling of observed regularities and their method of processing the experimental data were different. According to Kolb and Pospisil, the rate of conversion of the volatile components from a solid polymer into the gas phase can be assumed to be proportional to their concentration:

$$-\frac{\partial C}{\partial t} = kC \tag{3.8}$$

Thus, in the process of gas extraction of the polymer the content of the

3.3 DETERMINATION OF VOLATILE SUBSTANCES IN POLYMERS

volatile impurities must change over time according to the exponential law

$$C = C_0 e^{-kt} \qquad (3.9)$$

The authors[93] further assume that if the gas extraction is carried out carefully and for equal times, and equal portions of the head-space gas are introduced into the chromatograph, then the peak areas of a given volatile component in the chromatograms will follow the same (3.9) exponential law. In this case, the number of extractions accomplished after the first can be substituted for time t. In other words, the peak area of the volatile impurity after the ith stage of extraction (in the ith chromatogram of the gas phase) is related to the peak area in the first chromatogram according to

$$A_i = A_1 e^{-(i-1)k} \qquad (3.10)$$

With an infinitely large number of extractions, the sum of the areas of the individual peaks forming a decreasing geometric progression with $q = e^{-k}$ will be

$$\sum_{i=1}^{i=\infty} A_i = \frac{A_1}{1-q} \qquad (3.11)$$

With an infinite number of extractions, all volatile impurities will be converted into the gas phase and therefore the sum (3.11) will correspond to their total content in the analyzed sample. For polymer analysis there is no need to accomplish complete gas extraction; rather, it is sufficient to establish A_1 and q.

Before discussing the different methods used to accomplish this type of head-space analysis of the volatile impurities present in polymers, a few comments are necessary. The basis of the quantitative interpretation of the results of the so-called discontinuous gas extraction of polymers in the work of Kolb and Pospisil is equation (3.8) dealing with the rate of evaporation of volatile compounds. This is independent of the equilibrium conditions between the gas and the polymer. Furthermore, conversion of equation (3.9) to (3.10) is not strictly proven in the quoted paper.[93] However, the geometric progression following from (3.10) can be valid even without reference to the kinetic equation (2.8). If the process is accomplished under equilibrium conditions, then the geometric progression occurs as a result of the distribution law. Thus, the discussed method is based only on the dependence of (3.10) and it may be considered as an empirical relationship valid under specific experimental conditions for a large number of polymeric materials. This empirical relationship can be formulated simply as a linear dependence of the logarithm of the chromatographic peak area from the number of successive gas extractions. This was done by Suzuki et al.[94]:

$$\ln A_i = \ln A_1 + k - ki \tag{3.12}$$

The method of experimental determination necessary for the calculation of $k(q)$ depends on the degree of accuracy, simplicity, and the required speed of analysis.

The simplest (but not always the most accurate) way of calculating the total peak area according to (3.11) is the determination of k from the results of two repeated extraction. In this case, $k = \ln (A_1/A_2)$ and

$$\sum_{i=1}^{i=\infty} A_i = \frac{A_1^2}{A_1 - A_2} \tag{3.13}$$

Naturally, better results can be expected if one is not limited to two extractions. Here, data for a series of measurements are obtained and divided into two groups for the following computations, as shown in Figure 3.20. This method was recommended to calculate the average value of the numbers of extractions for each of the groups (m and n) and, correspond-

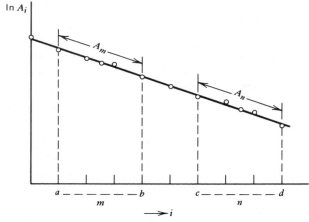

Figure 3.20. Determination of the average values of the number of extractions m and n, and the peak areas A_m and A_n.

3.3 DETERMINATION OF VOLATILE SUBSTANCES IN POLYMERS

ingly, the average values of the logarithms of the peak areas ($\ln A_m$ and $\ln A_n$).

$$k = \frac{\ln A_m - \ln A_n}{n - m} \qquad n > m \tag{3.14}$$

From the values obtained, the average value of k can be determined (3.14) as well as the corresponding (extrapolated) value of A. Thus, four extractions will yield (modifying the formula given by Kolb and Pospisil[93]):

$$\sum_{i=1}^{i=\infty} A_i = \frac{[(A_1 A_2)^7]^{1/8}}{[(A_1 A_2)^{1/4} - (A_3 A_4)^{1/4}](A_3 A_4)^{1/8}} \tag{3.15}$$

The computation of the total area corresponding to the content of volatile impurities is completed after calibration with standard mixtures. For the analysis of similar industrial polymer samples with similar impurity concentrations, only one determination of q is needed for the entire series. Individual samples can be extracted once to obtain A.

There are literature reports of data obtained with discontinuous gas extraction techniques in the determination of styrene in polystyrene,[93] of residual toluene in printed films of ethylene oxide in sterilized medical gloves,[93] and of water in insoluble polymers.[95] The latter case used the simplest method and employed one repeated extraction. The water content

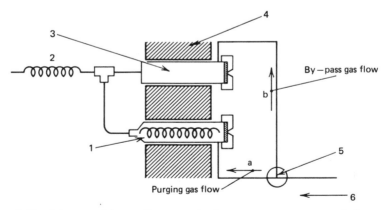

Figure 3.21. Schematic of the sampling system for discontinuous gas extraction. *1*, Gas extractor with the sample; *2*, chromatographic column; *3*, heated injector; *4*, thermostated injection block; *5*, three-way valve; *6*, carrier gas inlet.

in the polyethylene was 6.6–630 ppm, but in the polyamide it was present in quantities up to 1.9%.

The design of the apparatus used in the operations for discontinuous gas extraction is given in Figure 3.21. The size of the glass extraction tube is 11 × 85 mm. Changing the path of the carrier gas either to the extraction tube or directly into the column is accomplished manually using a three-way valve. A solenoid valve can be easily substituted for the three-way valve to automatically switch the timer or integrator. Therefore, the control of the duration of the extraction and the entire analysis is easily automated.

Figure 3.22 displays chromatograms of the vapor phase during four successive half-hourly extractions of a polystyrene sample (45 mg, 0.5 mm

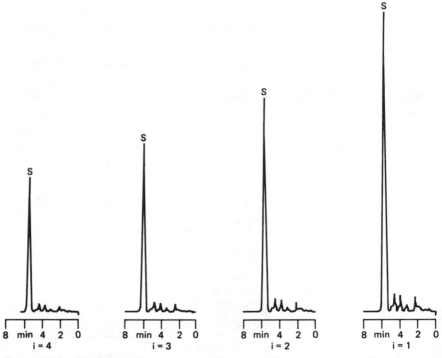

Figure 3.22. Head-space chromatograms of four consecutive gas extractions of a polystyrene sample. The peak marked S corresponds to the residual styrene monomer. Perkin-Elmer F-22 gas chromatograph. Column: 2 m × 0.25 in. o.d., containing 15% Carbowax 1500 on diatomaceous earth, 60-80 mesh. Column temperature: 110°C. Flame-ionization detector; attenuation: 100 × 16.

3.3 DETERMINATION OF VOLATILE SUBSTANCES IN POLYMERS

thick) at 150°C (purging time of the tube, 9 sec; styrene concentration, 620 ppm).

The vapor-phase analysis of adhesive tapes to determine the content of residual solvents[94] was completed by winding a 10 × 275 mm strip of the tape into a spiral shape around a drum. This was placed in the evaporation chamber with a free volume of 5 ml. The chamber, with a six-way valve, is located in a heating unit constructed of an aluminum alloy, the temperature of which is regulated with an accuracy up to ±0.5°C over a range up to 200°C. A temperature of 70°C is recommended for the analysis. The residual solvents (10–50 ppm of hexanes, cyclohexane, and methylcyclopentane) in tapes with an adhesive layer on a natural rubber base were determined with a relative standard deviation of not more than 8.3% (averaging approximately 6%).

The results obtained using the direct determination of volatile impurities released from polymers allow hope that this type of head-space analysis will acquire a much broader application.

Besides the direct determination of impurity concentrations in the polymer itself, the technique of head-space analysis is used successfully in hygienic studies of the polymers. The goal of these studies is the determination of the polymer characteristics as a source of environmental contamination. The knowledge of the concentrations of the volatile impurities in polymeric products and materials is insufficient to determine these characteristics. For the study the intensity and volatilization dynamics of harmful substances as they are released from the polymers into the atmosphere, all techniques of head-space analysis can be used. In many cases, the evolution of a gas is considerable, and direct gas-chromatographic analysis can be employed. Small air samples are taken from hermetically sealed containers containing the polymers (see, for instance, the studies on the evolution of gases from butadiene rubbers[96] and construction materials with poly(vinyl chloride)[97] or polystyrene base.[98] Very low concentrations can be determined by applying a preliminary concentrator like those used during the analysis of other substances.[99] Following this, thermal desorption[100]* or elution by solvents[101] occurs. In addition to the direct investigation of the gas evolved from polymers, head-space analysis is also applicable to the study of

---

*In this work, the method of equilibrium concentration in nonvolatile solution, described in detail in Chapter 4, was used.

the diffusion of harmful impurities into other media in contact with the polymers. The most tested medium is water. Head-space analysis was used in determining the volatile impurities in water extracted from granules of block polystyrene and a copolymer of styrene with methyl methacrylate.[102,103]

Special attention is given to the control of contamination of food products by impurities derived from polymeric packaging materials. An example of head-space analysis in this area is the work of Wilks and Gilbert.[104]

3.4 ANALYSIS OF FOOD PRODUCTS

In the investigation of food products, head-space analysis is particularly useful for the determination of

1. dangerous substances that either contaminate the raw food products or develop during storage
2. the volatile components of the food products that are responsible for their characteristic aroma and taste.

The American Chemical Society sponsored a special symposium on food and agriculture in which the modern aspects of head-space applications (in equilibrium and nonequilibrium conditions) for the analysis of foodstuffs were presented.[105] The scope of applications of head-space analysis for foodstuffs is quite broad. The general problems encountered include: sampling the vapor phase, quantitative analysis of the phase, concentration of the volatile components, and specific analytical problems regarding individual types of products. Most frequently head-space analysis is used for coffee, tea, cocoa, spices, fruits, beer, wine, and other alcoholic beverages.

A special application of head-space analysis is concerned with fruits and vegetables. This investigation encompasses the problems of storage, transport, and selection of the optimum packing methods, assuring proper ripening and the retention of taste qualities. A typical example of these investigations include the study of the composition of gases (O_2 and CO_2) over bananas packed in a polymer film of varying permeability.[106]

3.4 ANALYSIS OF FOOD PRODUCTS

The characterization of volatile components is limited by the qualitative analysis and identification discussed in Chapter 5. The content of volatile substances is rarely determined quantitatively, since it is not always necessary to separate individual components for comparative evaluation of foodstuff. Thus, it is sufficient to compare the quantities of the volatile fractions in the vapor phase under certain regulated conditions. An example is the investigation of the variation in the composition of the volatile aroma substances of coffee depending upon the packing[107] and roasting methods.[108] The total quantity of the volatile substances in an air sample taken above the ground and heated sample is determined by the total area of the chromatographic peaks. The dynamic change in the individual volatile components serves as the basis for determining the optimum modes of manufacture, storage, and product processing through study of the chemical processes involved.

An application of head-space analysis was found in the fermentation industry to determine secondary fermentation products—acetaldehyde, ethyl acetate, and fusel oil alcohols—with help of the semiautomatic analyzers[109,110] described in Chapter 2. The same instrumentation was used to determine the volatile carbonyl compounds in barley and malt.[111]

Head-space analysis of dangerous volatile impurities of foodstuffs is usually carried out to check for residual monomers and solvents leached from the polymer packing materials and containers. To this end, gas-chromatographic head-space analysis is much more accurate, more reliable, and more sensitive than chemical and organoleptic methods. Thus, the analysis of the vapor phase above cheese or cottonseed oil contained in polymer packing allows trace residual toluene to be detected immediately, whereas with sensory evaluations of these products, the toluene is detected only after three to four days storage in the polymer container. According to the data of Carlo Erba Co., its automatic accessory for head-space analysis (see Chapter 2) allows the detection of 0.05 ppm vinyl chloride in vodka in poly(vinyl chloride) containers.

Methanol is easily determined by head-space analysis as part of dangerous food impurities of different (nonpolymer) origin.[112]

Trichloroethylene is used to achieve a better extraction of carotenoids in the modern production of fruit juices. This demands the determination of residual quantities of the toxic extraction agent in the finished product. Direct head-space analysis using an electron-capture detector[113] allows such determinations to be conducted on a nanogram-per-liter scale. This eliminates systematic losses incurred by the methods based on a preliminary extraction of trichloroethylene by pentane.

3.5 DETERMINATION OF GASES IN SOLUTIONS

Head-space analysis is especially applicable to the determination of dissolved gases, since in most cases their solubility is low and conditions for achieving high sensitivity and selectivity are fairly favorable. The solubility and concentration of gas solutions are traditionally expressed in fractions or percentages of solvent volume. In such cases the volume of the dissolved gas is usually expressed under normal conditions (pressure 760 torr or 1.013 bar, temperature 0°C). The gas solubility coefficients expressed in volume percentages (Bunsen coefficients) represent 100 times the distribution coefficients of solutions saturated with pure gases under atmospheric pressure. Frequently the solubility of volumes of dissolved gases at a given temperature t (not 0°C) can be characterized by the "Ostwald coefficient," which is $(100t/273)\%$ higher than the Bunsen coefficients.

Normally one disregards the concentration dependence of the distribution coefficients of gases, which differ little in solubility, and uses data according to the Bunsen or Ostwald coefficients and partial pressures that are considerably less than atmospheric pressure. The concentration factor in the distribution law refers to nondissociated "free" molecules. However, one must be aware that gases having acid-base properties and, being subject to electrolytic dissociation, result in concentrations that do not represent the total composition of the given substance in solution.

For the calculations of head-space analysis of gases, literature data on solubility coefficients and Henry's constants are usually used. However, one must be attentive not only to the concentration and pressure units used, but also to the need to recalculate these values, because the calibration of the gas-chromatographic detectors is usually done in weight-volume concentrations. Table 3.4 gives solubility coefficients of the most important gases in water. These values are taken from the most recent and complete summary of existing data.[114]

As indicated in Chapter 1, the sensitivity of head-space analysis depends

Table 3.4 Solubility of Gases in Water—Values of the Ostwald Coefficient[a] (at 20°C)

Gas	Temperature, °C				
	10	20	25	30	40
Hydrogen	0.0203	0.0194	0.0191	0.0190	0.0189
Nitrogen	0.0196	0.0169	0.0159	0.0151	0.0140
Oxygen	0.0396	0.0334	0.0311	0.0292	0.0265
Carbon monoxide	0.0291	0.0249	0.0233	0.0221	0.0202
Carbon dioxide	1.238	0.937	0.828	0.739	0.605
Methane	0.0449	0.0367	0.0340	0.0316	0.0280
Ethane	0.0691	0.0514	0.0453	0.0405	0.0337
Ethylene	0.158	0.128	0.116	0.107	0.0925
Acetylene	1.363	1.108	1.013	0.935	0.816
Propane	0.0599	0.0424	0.0366	0.0322	0.0261
Propylene	—	0.216	0.181	0.152	0.107
Methylacetylene	2.424	1.846	1.671	1.537	1.315
Ammonia	406.6	342.2	312.7	285.2	236.0
Hydrogen sulfide	3.562	2.792	2.510	2.278	1.925
Sulfur dioxide	55.81	40.74	35.14	30.48	23.28

[a]Fraction of a volume of water at a partial pressure of 1 atm.

not only on the distribution coefficient, but also on the phase volume ratio. With low distribution coefficients the role of the latter factor becomes especially important, and during the analysis of dissolved gases the correct selection of phase volumes is frequently a determining factor. Figure 3.23 illustrates the effect of the phase volume ratio on the degree of sensitivity increase using head-space analysis as compared to the direct introduction of the analyzed solution into the chromatograph. Thus, for substances with $K = 10$ a change in the ratio V_G/V_L by two orders of magnitude (from 0.1 to 10) leads to a twofold change in the sensitivity of the analysis. But at $K = 0.1$ in the same range of phase volume ratio, the sensitivity (and the limit of detection) increases by 50 times.

Certain additional complications related to the volatility of the gases occur during the preparation of standard solutions used to calibrate the instruments and test the analytical methods. Known microgram quantities of hydrogen, oxygen, and nitrogen are assigned coulometrically by the electrolysis of solutions of potassium sulfide and hydrazine directly in stripping cells equipped with platinum electrodes. Saturated liquids outgassed by pure slightly soluble gases or gaseous mixtures of known composition are also used. The concentration of gases in the solutions is calculated according to Henry's law. The prepared standard solutions are transferred into analytical cells using tubes that prevent atmospheric contact so as to eliminate the loss of the gaseous components.

Traditionally, the determination of dissolved gases has frequently been

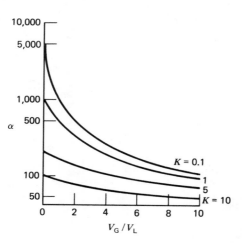

Figure 3.23. Sensitivity of head-space analysis (see Equation 1.43) vs. the relative volumes of the phases.

3.5 DETERMINATION OF GASES IN SOLUTIONS

completed using preliminary extraction in a vacuum. Various types of vacuum techniques are used. This method assumes that vacuum allows more rapid extraction of the gaseous components. Under laboratory conditions, common glass instrumentation (flasks, gas burets, and stopcocks) and the simplest mercury pumps, based on the principle of connecting flasks, are used. One of the standard methods recommended by the International Electrotechnical Commission for gas extraction from insulating oils[115] is illustrated in Figure 3.24. In these systems, for gases with sufficiently low distribution coefficients (to several units), the required degree of extraction (97–99%) is achieved in a small number of cycles. Under favorable conditions (high gas-liquid volume ratio) it can be limited to once lowering the level of the mercury, that is, developing a vacuum. For routine industrial measurements where a large number of samples are analyzed, more convenient instrumentation is made of metal and equipped with oil vacuum pumps. For example, according to reference 116, the vacuum treatment of a sample is accomplished using a special attachment placed directly in front of

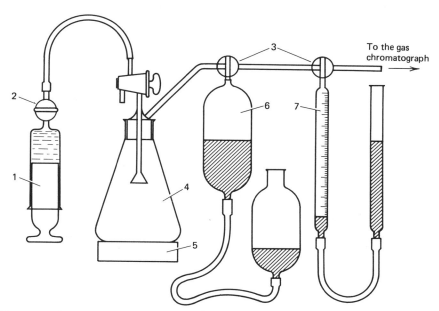

Figure 3.24. Schematic of a system for withdrawing dissolved gases from insulating oils with help of a vacuum pump. *1* Syringe with the liquid sample; *2*, two-way stopcock; *3*, three-way stopcocks; *4*, degasing flask; *5*, heater; *6*, mercury pump (Heisler or Topler pump); *7*, gas buret.

a gas chromatograph and having a system of gas valves. A liquid sample (approximately 1 ml) is introduced using a syringe through a tee fitted with rubber stoppers and connected to the carrier gas line. This system excludes an inflow of air. Gases dissolved in nonvolatile liquids are extracted at residual pressures less than 1 mm Hg while the analyzed sample is heated and magnetically stirred at 100° for 5 min. The discrepancy between the results of individual determinations does not exceed ±2.5%. The extraction devices for dissolved gases have been repeatedly patented and produced on a limited scale. As an example of a device produced for outgassing a fluid, UDZh-64 was manufactured at the Dzerzhinsk (U.S.S.R.) branch of O.K.B.Automatiki in 1977 (Fig. 3.25). It is a quite complex unit weighing 80 kg that works in combination with the Tswett 100 series gas chromatographs. The dissolved gases are extracted under a vacuum of approximately 3 mm Hg at temperatures up to 200°C and then purged by the carrier gas into the chromatographic column. One to two milliliters of the outgassed liquid is introduced (using a syringe or buret) into a U-shaped glass pipet-extractor of 30-ml volume. Two sample analyses can be accomplished in parallel on different columns. This method is necessary for complex gas mixtures that do not separate on one stationary phase.

Various devices for gas extraction from liquids using vacuum are quite popular. It should be noted that the utilization of these systems for gas-

Figure 3.25. Model UDZs-64 instrument for the outgassing of liquids. *1*, Model TR-15 oven for the extraction columns; *2*, Model BPG-76 unit for gas preparation and vacuum regulation.

chromatographic head-space analysis is not compulsory and in many cases becomes cumbersome. Here, the general examples of HSA, whether static or dynamic as discussed in Chapter 1, can be applied effectively and easily.

One of the early examples of extraction using an inert gas flow involves the determination of CO_2, O_2, N_2, CH_4, and CO in water. Such determinations have a detection limit of 0.3 ppm. The glass cell used for this purpose is shown in Figure 3.26. One to two milliliters of analyzed water is introduced by syringe through a rubber septum. A purge of helium (50 ml/min), having passed through a porous glass disk, carries the dissolved gaseous components into a dryer and then into the chromatographic column.[117] The peaks obtained under stripping conditions are somewhat indistinct. Thus, quantitative analysis must be accomplished using peak areas instead of peak heights.

When the liquid volumes exceed 3 ml, stripping is prolonged and the broadening of the chromatographic peaks becomes so extensive that direct introduction into the gas chromatograph becomes impractical. Further increasing the sample volume to lower the detection limit requires, in addition to stripping, an intermediate gas trap. In a helium stripping device (60 ml/min), the extraction of dissolved hydrogen, nitrogen, and oxygen

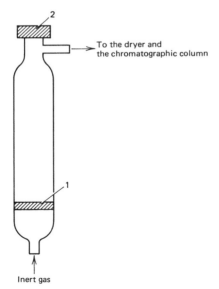

Figure 3.26. Schematic of glass cell used for the degassing of aqueous solutions by inert gas flow. 1, Porous disk with large pores; 2, rubber septum.

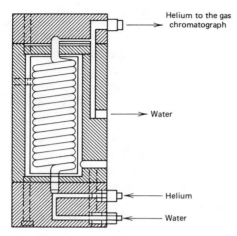

Figure 3.27. Schematic of a device for the continuous stripping of dissolved gases from flowing water; system for low water flow rates.[119]

from a 30-ml water sample utilizing followed by the intermediate trapping of the gases by molecular sieves* was completed in 20 min.[118] The detection limit of hydrogen was 2×10^{-9} (of the mass of sample) and the oxygen and nitrogen 10^{-8} (of the mass of sample). In the concentration range of 0.1–1.0 ppm, an accuracy of ±10% is easily obtained.

Continuous analysis of dissolved gases requires devices that complete stripping in the flow of the liquid. The construction of such equipment differs considerably depending on the flow rates of the gas purge and the liquid. At low flow rates (up to 10 ml/min) the stripping can be completed in a vertical spiral cylinder constructed of glass tubing, 5.3 m in length and 2.5 mm i.d. (Fig. 3.27).[119] Flows of water and pure helium are supplied simultaneously at the lower end of the spiral. At the exit of the spiral the dissolved quantities (1 ml/kg of water) of hydrogen, oxygen, methane, carbon monoxide, and carbon dioxide are practically completely transferred into the gas phase and enter the chromatograph with the helium. The device shown in Figure 3.28 is designed for higher flow rates such that the basic atmospheric gases are extracted from water at flow rates up to 100 ml/min.[120] The high efficiency of this compact gas extractor is achieved by the application of the principle of counterflow and a series of rotating disks partially submerged in water. A thin film of moisture, entrained by the large

---

*Four milliliters of powder, obtained by grinding Linde molecular sieve 5A tablets with subsequent sieving through 30/60 mesh, are cooled to liquid nitrogen temperatures during trapping with desorption at room temperature.

3.5 DETERMINATION OF GASES IN SOLUTIONS

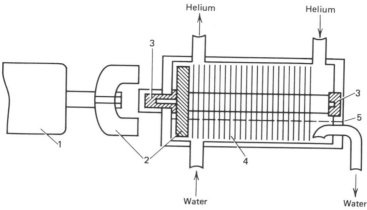

Figure 3.28. Schematic of a device for the continuous stripping of dissolved gases from flowing water; system for high water flow rates.[120] *1*, Electromotor; *2*, permanent magnets; *3*, Teflon bearings; *4*, rotating disks; *5*, optimum water level.

surface area of these disks during rotation, quickly establishes equilibrium with the gas phase. The construction of the system allows independent regulation of the water and gas flows. The ratio of these flow rates determines the concentration ratio of the analyzed gaseous components in the entering water and exiting gas phase.

Besides the determination of atmospheric gases in natural waters, the determination of gaseous hydrocarbons in electrical insulation oils and of hydrogen in boiler water represent important problems. Hydrogen occurs in the water of large steam boilers as one of the final products of alkaline, carbonate, steam, and water corrosion. Data on hydrogen concentration serve as an indication of the degree of pipe corrosion in the boiler and the necessity of repairs to prevent breakdown. Hydrogen solubility in water at 20°C and atmospheric pressure is 16.3 mg/kg. Therefore, the detection limit needed (approx. 0.1 mg/kg) can be reached with a liquid–gas volume ratio V_L/V_G in the order of 15. For such analyses, instrumentation[121] developed by the All-Union Energy Institute bubbles 80 ml of air through 1.2 liters of water using a membrane microcompressor. Equilibrium is established in 30–40 min. At that point several milliliters of the vapor phase are removed with a syringe and introduced into a chromatograph. Due to the problems of corrosion of steam boilers, it is necessary to also check the amount of dissolved oxygen and other gases in the water. The continuous helium stripping instrument was developed for this purpose.[119] This device, combined with a gas chromatograph using a helium ionization detector, allows the determination of the amounts of dissolved hydrogen, oxygen, methane, carbon monoxide and dioxide at the level of tenths of a milliliter per liter of water. The standard deviation is approximately 4% (except for CO and CO_2).

Ecological chemistry and environmental protection comprise important applications areas for the analysis of trace gases dissolved in water. Such investigations include a detailed study of the depth dependence of nitrogen oxide, Freon-11 ($CFCl_3$), and Freon-12 (CF_2Cl_2) in seawater.[122] These data are of considerable interest in connection with the accumulation of freons in the atmosphere. Such accumulation endangers the ozone layer that protects the living organisms of the earth from destruction by ultraviolet radiation. Freons are fairly soluble in water, and their concentration at depths up to 2–3 km drops to 10^{-11} g/liter. Therefore, considerably large seawater samples (0.5 liter) were vacuum-extracted to determine their freon content.

The determination of gases in insulation oils by head-space analysis is a developing area of some interest. Studies conducted in 1960–1970 in many countries with highly developed electrical energy show that determination of trace dissolved gases in transformer oil can be used as a reliable and effective method for detecting and diagnosing power transformer defects that develop during operation. This method of monitoring the condition of high-voltage power transformers is of considerable economic importance due to the preventive abilities regarding serious damage and the timely elimination of potential damages in their early stages. The gases develop in the transformers due to the effect of heat and electrical discharges on the insulation. The disintegration of the cellulose in the insulation and electrotechnical pressboard leads to a release of carbon monoxide and dioxide* into the transformer oil. Also, during the pyrolysis of solid insulation and electroinsulating oils, methane and ethylene and their homologues develop. At temperatures over 600°C or with arc discharges, acetylene results. Small quantities of gases are released slowly during the natural aging of the insulation in normally operating transformers. However, statistical analysis of a large number of installations in various countries indicates that the content of gaseous components in oils of nondefective transformers does not, as a rule, exceed the limits given in Table 3.5. If the normal mode of operation is disturbed (as in overheating and electrical discharges), the

---

*The CO_2 content in the oil of operating transformers can reach 1% by volume.

Table 3.5 Limiting Concentrations of Gases in the Oil of Operative Transformers (Volume Percents)

Gases	Switzerland[123]	USSR[124]	France[125]	GFR[126] Time of Exploitation	
				Up to 5 yr.	5–10 yr.
Hydrogen	0.02	0.002	0.01	0.005	0.01
Methane	0.005	0.01	0.04	0.005	0.01
Ethane	0.0015	0.01	0.02	0.01	0.02
Ethylene	0.006	0.03	0.007	0.01	0.03
Acetylene	0.0015	0.001	0.0001	0.0015	0.003
Propane	—	—	0.02	0.005	0.01
Propylene	—	—	0.01	0.01	0.03
Carbon monoxide	0.1	—	0.08	0.03	0.05
Carbon dioxide	1.1	—	1.01	0.3	0.5

insulation is destroyed. Here, the gas content rapidly increases and exceeds by manyfold the reference concentration limits. The simultaneous excess of the two limiting values as given in Table 3.6 must be considered an indication of damage.[123] In the Soviet literature,[127] special attention has been given to the presence of acetylene (which develops only at high temperatures and strong electrical discharges) and the accompanying increase in the carbon dioxide and methane content. It might be noted that the low gas content in insulation oil does not guarantee the absence of damage. In large transformers the total quantity of oil reaches tens of tons. Thus, while the concentration of the decomposition product near the actual damage site may be high, it will be relatively low further away from the damage site and in average will not reach the specified limits. The damage was frequently detected during the inspection of transformers in which the gas content in the oil did not reach the indicated limiting values.[125] In relation to this, the most important criterion for the presence of damage is the rapid increase of the concentration of dissolved gases over certain time intervals as opposed to the absolute concentrations of the gases.[126,130]

More reliable and detailed information on the condition of operating transformers is obtained when the concentration *ratios* of certain dissolved gases are used as a diagnostic tool. The diagnostic system recommended[128] by the International Electrotechnical Commission is based on the determination of the relative concentrations of five dissolved gases in transformer oil (methane, ethane, ethylene, acetylene and hydrogen) and is utilizing three characteristic concentration ratios: C_2H_2/C_2H_4, CH_4/H_2, and C_2H_4/C_2H_6. Normal aging of the insulation (without defects) causes the first and last of these ratios not to exceed 0.1, while the second can vary from 0.1 to 1. When partial discharges through a gas-filled cavity occur, the relative content of methane decreases. With a spark discharge, the insulation is ruptured and an intense electrical arc develops. Here, the relative content of acetylene (and ethylene) increases. Local overheating is accompanied by an increase in the methane content, which at temperatures over 150°C exceeds the hydrogen content, and at temperatures over 300° the ethylene concentration exceeds the ethane concentration. Thus, it is possible to detect and distinguish 8–12 types of characteristic defects in power transformers.[128]

The method of evaluating the operation of transformers from the results of gas analysis is fairly sensitive and can give important information as to the order in which the transformers should be overhauled. Head-space analysis is the most suitable solution to this problem. Methods including complete as

3.5 DETERMINATION OF GASES IN SOLUTIONS

well as partial extraction of the dissolved gases were the most practical. A method of head-space analysis was mentioned in one of the early works concerning gas-chromatographic control in oil transformer manufacturing.[129] However, its discussion and coefficient calculations remain unpublished.

A method of analysis[125] including the extraction of gases by the carrier gas in a bubbler (Fig. 3.29) mounted directly in front of the chromatographic column was developed in the French Central Laboratory of Electro-Industry. For the determination of hydrocarbons and carbon monoxide and dioxide, a 0.25-ml oil sample is necessary. The extraction is completed in several seconds. Under such conditions the detection limit for hydrogen using a thermal conductivity detector is 4×10^{-6} mole/liter, which is unacceptably low. Therefore, hydrogen is determined in separate oil samples of 5–10 ml, which are placed in differently designed extracting cells (Fig. 3.30). The detection limit of gaseous hydrocarbons in transformer oil by this method is 0.5 ppm (by volume), and that of hydrogen is 2 ppm. The method of incomplete extraction appears among the recommended methods of analysis chosen by the International Electrotechnical Commission.[115] This analysis is conducted using fairly complex glass instrumentation and high liquid–gas volume ratios. Vacuum is formed by lowering the level of mercury. Thus, any unextracted quantities of gases, which by this method are small, are taken into consideration by using distribution coefficient values obtained by equation (1.20).

Methods based on complete or partial extraction of the dissolved gases[131] are also used in the USSR. For complete vacuum extraction the above-described equipment[116] (UDZh 64) is used. The sensitivity for methane is 5×10^{-3}% (volume) in a 2–5-ml oil sample. Partial extraction[132] has achieved wide utilization but has been criticized.[131] The impact of the criticism lies in

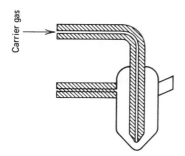

Figure 3.29. Bubbler for the extraction of gases dissolved in transformer oil.

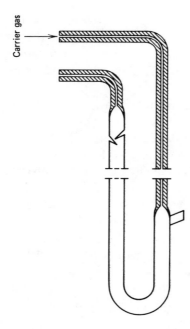

Figure 3.30. Cell for the extraction of trace amounts of hydrogen dissolved in transformer oil.

the fact that it gives lower results, which vary depending on the extraction time. In the final modification of the partial extraction method,[124] additional 100-ml sealed glass syringes are used. A syringe holds 30 ml of oil, and the vacuum is formed by pulling the plunger up to the 100-ml mark. Then the extracted gas is supplied by a plunger into a gas chromatographic flowmeter. Incomplete gas extraction under these conditions necessitates the introduction of empirically computed coefficients of 1.1 for methane, 2.2 for carbon dioxide and ethylene, 1.7 for acetylene, and 2.7 for ethane. The quantity of each gas detected in the vapor phase is multiplied by the coefficient to determine its content in the analyzed oil. Hydrogen is extracted almost completely, that is, the extraction coefficient for hydrogen is equal to 1. Most of the computed coefficients given differ considerably from the values $(K + V_G/V_L)$ that were used for the calculation under strict equilibrium conditions. The possibility of error, related to nonequilibrium conditions in this case, remains.

REFERENCES

1. B. Kolb, H. Krauss, and M. Auer, *Applications of Gas Chromatographic Head-Space Analysis*, Vol. 21, Bodenseewerk Perkin-Elmer GmbH, Überlingen 1978.
2. I. L. Butaeva, Razrabotka Gazokhromatograficheskogo Opredeleniya Mikroprimesei Letuchikh Sernistykh Veshchestv v Stochnykh Vodakh Sul'fatno-tsellyuloznogo Proizvodstva Metodom "Analiza Ravnovesnogo Para," Kand. Diss., Leningrad Technical University, 1978, 143 pp.
3. I. L. Butaeva, V. V. Tsibulskii, A. G. Vitenberg, and M. D. Inshakov, *Zh. Analit. Khim.*, **28**, 337 (1973).
4. A. G. Vitenberg, I. L. Butaeva, L. M. Kuznetsova, and M. D. Inshakov, *Zh. Prikl. Khim.*, **49**, 1476 (1976).
5. A. G. Vitenberg, I. L. Butaeva, L. M. Kuznetsova, and M. D. Inshakov, *Anal. Chem.*, **49**, 128 (1977).
6. C. McAuliffe, *Chem. Technol.*, **1**, 46 (1971).
7. C. McAuliffe, U.S. Patent 3,759,086, (1973).
8. B. V. Ioffe, B. V. Stolyarov, and S. A. Smirnova, *Zh. Analit. Khim.*, **33**, 2196 (1978).
9. S. A. Smirnova, Novye Vozmozhnosti Gazokhromatograficheskogo Opredeleniya Letuchikh Organicheskikh Veshchestv v Vodnykh Rastvorakh Metodom Analiza Ravnovesnogo Para, Kand. Diss., Leningrad State University, 1977, 124 pp.
10. J. Drozd, J. Novák, and J. A. Rijks, *J. Chromatogr.*, **158**, 471 (1978).
11. M. Gottauf, *Z. Anal. Chem.*, **218**, 175 (1966).
12. V. E. Stepanenko and Z. M. Golovina, *Zh. Analit. Khim.*, **30**, 890 (1975).
13. E. S. K. Chian, P. P. K. Kuo, W. J. Cooper, W. F. Cowen, and R. C. Fuentes, *Environmental Sci. Technol.*, **11**, 282 (1977).
14. V. V. Naumova, N. I. Rakita, M. L. Sazonov, and A. A. Zhukhovitskii, USSR Patent 258,713, *Byull. Izobr.*, No. 1, 1970.
15. M. L. Sazonov, A. A. Zhukhovitskii, N. I. Rakita, and V. V. Naumova, *Tr. Vses. Nauchn. Issled. Geol. Neft. Inst.*, **112**, 11 (1973).
16. B. K. Krylov, A. S. Blinov, M. L. Sazonov, V. V. Naumova, and N. I. Rakita, USSR Patent 549,737, *Byull. Izobr.* No. 9, 1977.
17. Z. Voznáková, M. Popl, and M. Berka, *J. Chromatogr. Sci.*, **16**, 123 (1978).
18. P. P. K. Kuo, E. S. K. Chian, F. B. DeWalle, and J. H. Kim, *Anal. Chem.*, **49**, 1023 (1977).
19. T. A. Bellar and J. J. Lichtenberg, *J. Am. Water Works Assoc.*, **66**, 739 (1974).
20. R. van Wijk, *J. Chromatogr. Sci.*, **8**, 418 (1970).
21. A. Zlatkis, H. A. Lichtenstein, and A. Tishbee, *Chromatographia*, **6**, 67 (1973).

22. W. Bertsch, R. C. Chang, and A. Zlatkis, *J. Chromatogr. Sci.*, **12**, 175 (1974).
23. W. D. Snyder, "Automated Analysis of Volatile Organic Compounds in Water," Hewlett-Packard Co. Tech. Paper **GC-71**, 1978.
24. K. Grob and F. Zürcher, *J. Chromatogr.*, **117**, 285 (1976).
25. J. Novák, J. Zluticky, V. Kubelka, and J. Mostecky, *J. Chromatogr.*, **76**, 45 (1973).
26. S. P. Wasik, *J. Chromatogr. Sci.*, **12**, 845 (1974).
27. A. G. Vitenberg, B. V. Stolyarov, and S. A. Smirnova, *Vestn. Leningr. Gos. Univ.*, **1977** (16), 132.
28. N. C. Jain and R. H. Cravey, *J. Chromatogr. Sci.*, **10**, 263 (1972).
29. R. H. Cravey and N. C. Jain, *J. Chromatogr. Sci.*, **12**, 209 (1974).
30. A. S. Curry, *J. Chromatogr. Sci.*, **12**, 529 (1974).
31. J. A. Hancock, F. L. Mill, and J. R. Miles, *Clin. Toxicol.*, **4**, 217 (1971).
32. P. K. Wilkinson, J. G. Wagner, and A. J. Sedman, *Anal. Chem.*, **47**, 1506 (1975).
33. Metodicheskoe Pis'mo po Provedeniyu Raschetov, Svyazannykh s Ekspertizoi Alkogol'noi Intoksikatsii na Trupe, Min. Zdravookhr. RSFSR. Chelyabinsk, 1974, 26 pp.
34. B. L. Glendening and R. A. Harvey, *J. Forensic Sci.*, **14**, 136 (1969).
35. N. C. Jain, *Clin. Chem.*, **17**, 82 (1971).
36. V. J. Perez, T. J. Cicero, and B. A. Bahn, *Clin. Chem.*, **17**, 307 (1971).
36a. R. N. Harger, B. B. Raney, E. G. Bridwell and M. F. Kitchel, *J. Biol. Chem.*, **183**, 197 (1950).
37. L. R. Goldbaum, T. J. Domansci, and E. L. Schloegel, *J. Forensic Sci.*, **9**, 63 (1964).
38. M. L. Luckey, *J. Forensic Sci.*, **16**, 120 (1971).
39. D. Jentzsch, H. Krüger, G. Lebrecht, G. Dencks, and J. Gut, *Z. Anal. Chem.*, **236**, 96 (1968).
40. G. Machata, *Microchim. Acta*, **1964**, 26.
41. G. Machata, *Blutalkohol*, **4**, 252 (1967).
42. G. Machata, *Clin. Chem. Newsletter*, **4**(2), 29 (1972).
43. G. A. Brown, D. Neylan, W. J. Reynolds, and K. W. Smaldon, *Anal. Chim. Acta*, **66**, 271 (1973).
44. K. W. Smaldon and G. A. Brown, *Anal. Chim. Acta*, **66**, 285 (1973).
45. A. F. Rubtsov and E. M. Salomatin, Sbornik Trudov NII Sudebnoi Psikhiatrii "Aktual'nye Voprosy Sudebno-Meditsinskoi Ekspertizy Trupov" Moscow., 1977, 116 pp.
46. G. Machata, *Z. Rechtsmed.*, **75**, 229 (1975).
47. G. Hayck and H. P. Terfloth, *Chromatographia*, **2**, 309 (1969).
48. Perkin-Elmer Multifract F-40 Gas Chromatograph for Blood Alcohol Analyses, Techn. Paper 927, 1973.

REFERENCES

49. D. Jentzsch, H. Krüger, and G. Lebrecht, *Angew. Gas-Chromatogr.*, Vol. 10E, Bodenseewerk Perkin-Elmer GmbH, Überlingen, 1967, 21 pp.
50. H. J. Battista, *Z. Rechtsmed.*, 72, 278 (1973).
51. G. Machata, *Blutalkohol*, 7, 345 (1970).
52. G. Machata and L. Prokop, *Blutalkohol*, 8, 349 (1971).
53. Head-Space Sampler HS-6 for Gas Chromatography, Perkin-Elmer Corp., Norwalk, Conn.; Publ. 1738 (1978).
54. H. Hachenberg and A. P. Schmidt, *Gas Chromatographic Headspace Analysis*, Heyden, London, 1977, 125 pp.
55. B. Kolb, *J. Chromatogr.*, 122, 553 (1976).
56. A. A. Bruk and V. I. Pomerantsev, *Sudebno-Med. Ekspertiza, Min. Zdravookhr. SSSR.*, 18(2), 44 (1975).
57. V. I. Pomerantsev, A. A. Bruk, and N. A. Gosteva, *Sudebno-Med. Ekspertiza, Min. Zdravookhr, SSSR.*, 18(4), 32 (1975).
58. J. Lindner and H. Weichardt, *Zentralbl. Arbeitsmed. Arbeitsschutz*, 22, 323 (1972).
59. T. Kojima and H. Kobajashi, *Japan. J. Leg. Med.*, 27, 255 (1973).
60. T. G. Field and J. B. Gilbert, *Anal. Chem.*, 38, 628 (1966).
61. J. Ruhnke, A. Eggert, and H. Huland, *Chromatographia*, 7, 55 (1974).
62. Metodicheskie Ukazaniya ob Obnaruzhenii i Opredelenii Atsetona v Trupnom Materiale Metodom Gazozhidkostnoi Khromatografii, Min. Zdravookhr. SSSR, Moscow, 1978.
63. R. Bonnichsen, A. C. Maehly, and M. Moeller, *J. Forensic Sci.*, 11, 186 (1966).
64. R. Bonnichsen and A. C. Maehly, *J. Forensic Sci.*, 11, 414 (1966).
65. Metodicheskie Ukazaniya ob Obnaruzhenii i Opredelenii 1,2-dikhloretana v biologicheskom Materiale Metodom Gazozhidkostnoi Khromatografii, Min. Zdravookhr. SSSR, Moscow, 1978.
66. S. Natelson and R. L. Stellate, *Microchem. J.*, 9, 245 (1965).
67. V. F. Pomerantsev, *Sb. Vopr. Kriminal., Sudebnoi Ekspertiza i. Kriminal.*, 1971 (6), 71.
68. D. J. Blackmore, *Analyst*, 95, 439 (1970).
69. G. Triebig, *Vorträge zum 2. Internationalen Colloquium Über die Gaschromatographische Dampfraumanalyse in Überlingen*, October 1978.
70. J. Lindner and H. Weichardt, *Z. Anal. Chem.*, 267, 347 (1973).
71. GOST15820-75. Metody Opredeleniya Ostatochnykh Monomerov Stirola, α-Metilstirola, Akrilonitrila, Metilmetakrilata i Nepolimerizuyushchikhsya Primesei Etilbenzola i Izopropilbenzola v Polistirol'nykh Plastikakh s Pomoshch'yu Gazovoi Khromatografii.
72. DIN 53741-71: Prüfung von Kunststoffen. Bestimmung flüchtiger aromatischer Kohlenwasserstoffe in Polystyrol. Gaschromatographische Verfahren.
73. Mezhdunarodynyi Standart 2561-74 (ISO). Polistirol: Opredelenie Ostatochnogo Monomera Metodom Gazovoi Khromatografii.

74. L. Rohrschneider, *Z. Anal. Chem.*, **255**, 345 (1971).
75. M. Swiatecka and H. Zowall, *Polymery tworzjwa wielkosz*, **1**, 33 (1975).
76. J. Puschmann, *Angew. Makromol. Chem.*, **47**, 29 (1975).
77. W. R. Eckert, *Fette, Seife, Anstrichmittel*, **71**, 319 (1975).
78. R. J. Steichen, *Anal. Chem.*, **48**, 1398 (1976).
79. J. Gilbert, J. R. Startin, and M. A. Wallwork, *J. Chromatogr.*, **160**, 127 (1978).
80. G. DiPasquale, G. Dilorio, and T. Capocioli, *J. Chromatogr.*, **160**, 133 (1978).
81. L. A. Zagar, *J. Pharm. Sci.*, **61**, 1801 (1972).
82. B. Kolb, *J. Chromatogr.*, **122**, 553 (1976).
83. H. Hachenberg, *Angew. Gas-chromatogr* Bodenseewerk Perkin-Elmer GmbH. Überlingen, **25**, 5 (1975).
84. A. R. Berens, *Polymer Prep.*, **15**, 197 (1974).
85. A. R. Berens, *Polymer Prep.*, **15**, 203 (1974).
86. A. R. Berens, *J. Appl. Polymer Sci.*, **19**, 3169 (1975).
87. N. A. Tarasova, V. D. Feofanov, and V. E. Gul', *Gigiena i Sanit*, **1971** (11), 114.
88. V. D. Feofanov, N. F. Tolikina, and O. N. Belyatskaya, "Gigiena Primeneniya Polimernykh Materialov i Izdelii iz Nikh," Kiev, 1969, p. 518.
89. S. J. Romano, J. A. Renner, and P. M. Leitner, *Anal. Chem.*, **45**, 2327 (1973).
90. ASTM F-151-72, Standard Test Method for Residual Solvents in Flexible Barrier Materials, Amer. Soc. for Testing & Materials, Philadelphia, Pa.
91. J. T. Davies and J. R. Bishop, *Analyst*, **96**, 55 (1971).
92. S. Cohadzic, *Farbe Lack*, **79**, 314 (1973).
93. B. Kolb and P. Pospisil, *Chromatographia*, **10**, 705 (1977).
94. M. Suzuki, S. Tsuge, and T. Takeuchi, *Anal. Chem.*, **42**, 1705 (1970).
95. B. Kolb and P. Pospisil, *Applications of Gas Chromatographic Head-Space Analysis*, Vol. 19, Bodenseewerk Perkin-Elmer GmbH, Überlingen, (1978).
96. A. V. Vikhmentsev, L. V. Timofeeva, and Yu. A. Novikov, in *Mediko-Tekhnicheskie Problemy Individual'noi Zashchity Cheloveka*, Vol. 10, Meditsina, Moscow, 1972, p. 120.
97. A. P. Filippov, V. K. Komlev, and V. V. Mal'tsev, *Gigiena i Sanit.* **1972** (6), 67.
98. G. I. Smirnova, V. V. Mal'tsev, V. K. Komlev, V. A. Plakhov, and V. Yu. Gvil'dis, *Gigiena i. Sanit.*, **1972** (3), 58.
99. G. I. Rudenko, Sanitarno-Khimicheskaya Otsenka Polimernykh Stroitel'nykh Materialov na Osnove Sinteticheskikh Kauchukov i Polistirola s Pomoshch'yu Metoda Gazo-Zhidkostnoi Khromatografii, Avtoreferat Kand. Diss. Moscow, 1978, 24 pp.
100. V. V. Mal'tsev, G. I. Smirnova, and A. F. Guk, *Plast. Massy*, **1975** (5), 35.
101. Yu. G. Shirokov, Yu. S. Drugov, G. V. Murezoeva, and G. L. Petunovskaya, *Gigiena i. Sanit.*, **1976** (1), 50.
102. M. A. Markelov and È. I. Semenenko, *Plast. Massy*, **1973** (8), 65.

103. M. A. Markelov and È. I. Semenenko, *Plast. Massy*, **1976** (1), 57.
104. R. A. Wilks and S. G. Gilbert, *Food Technol.*, **23**, 47 (1969).
105. G. Charalambous (Ed.), *Analysis of Food and Beverages. Head-space Techniques*, Academic Press, New York, 1973, 394 pp.
106. H. Daun, S. G. Gilbert, Y. Ashkenazi, and Y. Henig, *J. Food Sci.*, **38**, 1247 (1973).
107. V. N. Simonova and T. Ya. Solov'eva, *Izv. Vuzov. Pishchevaya Tekhnol.*, **1978** (4), 76.
108. T. Ya. Solov'eva and V. N. Simonova, *Izv. Vuzov. Pishchevaya Tekhnol.*, **1977** (0), 165.
109. B. Mandle, F. Wullinger, W. Binder, and A. Piendl, *Brauwissenschaft*, **22**, 477 (1969).
110. B. Wagner and G. Baron, *Monatsschr. Brauerei*, **24**, 225 (1971).
111. B. Wagner, *Monatsschr. Brauerei*, **24**, 28 (1971).
112. M. A. Bertolaccini and A. Bicca, *Boll. lab. chim. provinc.*, **23**, 437 (1972); *R.Zh.Khim.* 1G275 (1974).
113. B. Kolb and M. Auer, *Applications of Gas Chromatographic Head-Space Analysis*, Bodenseewerk Perkin-Elmer GmbH, Überlingen, 1978, 22 pp.
114. E. Wilhelm, R. Battino, and R. J. Wilcock, *Chem. Rev.*, **77**, 219 (1977).
115. IEC Standard, Publ. 567: "Guide for the Sampling of Gases and of Oil from Oil-Filled Electrical Equipment and for the Analysis of Free and Dissolved Gases," Geneva, 1977, 51 pp.
116. E. N. Stern, R. A. Lipstein, and V. V. Kulikova, *Khim. i Tekhnol. Topliva i Masel*, **1971** (3), 53.
117. J. W. Swinnerton, V. J. Linnenbom, and C. H. Cheek, *Anal. Chem.*, **34**, 483 (1962).
118. A. Tolk, W. A. Langerak, A. Kout, and D. Börger, *Anal. Chim. Acta*, **45**, 137 (1969).
119. J. A. J. Walker and E. D. France, *Analyst*, **94**, 364 (1969).
120. D. D. Williams and R. R. Miller, *Anal. Chem.*, **34**, 657 (1962).
121. A. A. Avdeeva, *Teploenergetika*, **1966** (4), 88.
122. A. Hahne, "Messung atmosphärischer Spurengase im Meerwasser," Institut für Chemie, Kernforschungsanlage Jülich GmbH, 1977.
123. E. Dornenburg and W. Strittmatter, *Brown Boveri Mitt.*, **61**, 238 (1974).
124. K. F. Stepanchuk, G. S. Klimovich, A. V. Sotnikova, I. V. Yacheiko, and M. A. Smirnov, *Izv. Vuzov. Energetika*, **1978** (7), 44.
125. M. Thibault and J. Rabaud. *Rev. gen. l'Electr.*, **84**, 81 (1975).
126. R. Miller, H. Schliesing, and K. Soldner, *Elektrizitätswirtschaft*, **73**, 683 (1974).
127. M. A. Smirnov and N. P. Fufurin, *Tr. Vses. Nauchn. Issled. Inst. Elektr.*, **49**, 24 (1976).

128. R. R. Rogers, *IEEE Trans. Electr. Insul.*, **13**, 349 (1978).
129. E. Dornenberg and O. E. Gerber, *Brown Boveri Mitt.*, **54**, 104 (1967).
130. M. S. Zuzak, V. L. Talover'ya, and E. N. Mishchenko, *Elektr. Stantsii*, **1978** (6), 75.
131. M. S. Zuzak and T. P. Mal'tseva, *Elektr. Stantsii*, **1978** (7), 85.
132. M. A. Smirnov, G. K. Kolobaev, and N. P. Fufurin, *Elektr. Stantsii*, **1973** (8), 57.

CHAPTER FOUR

Equilibrium Concentration of Impurities of Gases (Reverse Head-Space Analysis)

4.1 CHARACTERISTICS AND BASIC TYPES OF REVERSE HEAD-SPACE ANALYSIS

Modern requirements concerning the sensitivity of the methods for the determination of the amount of impurities present in gases, especially in air, are so stringent that in many cases direct introduction of the investigated gas into the gas chromatograph is insufficient to attain the required threshold sensitivity. Thus, while the direct analysis of gases using flame-ionization detector allows the determination of impurities at levels of 1–10 mg/m^3, the maximum allowable concentrations of organic substances in air is lower by one to three orders of magnitude. In order to attain the required analytical sensitivity, cryogenic or adsorption methods of concentration as well as the collection of substances and their accumulation in solutions must be used. These traditional methods are based upon the total extraction of the concentrated substance. As such they possess a series of shortcomings related to the necessity of water removal, prevention of the passage of the analyzed substances through the trap and so on. The majority of shortcomings in the traditional methods of concentration can be eliminated if the principle of equilibrium concentration is used instead of the tenet of total adsorption. In this case, the gas sample is first brought to thermodynamic equilibrium with a liquid phase, which subsequently is subjected to analysis.* Such a method in essence is the reverse of HSA methodology. It can be used successfully with favorable values of the distribution coefficients and with the maintenance of certain conditions resulting from the theory of equilibrium concentration. The method has two variations, each having specific characteristics regarding the technical design of the process as well as some different possibilities and limitations: using either a practically nonvolatile or a volatile liquid for the concentration of the impurities.

Equilibrium concentration in nonvolatile liquids (stationary phases for gas chromatography) was first suggested in 1965[3] and used in the works of

---

*The determination of impurity concentrations in gases, based upon the principle of phase equilibrium, was first used before the development of gas chromatography for the continuous monitoring of the concentration of carbon dioxide in air.[1,2] The method is based upon the saturation of an aqueous solution of sodium bicarbonate with the analyzed air sample. The pH of the solution, which depends upon the concentration of the carbon dioxide in the equilibrium gas, is measured. The method is now only of historical interest, because its sensitivity is low, only ~0.02%. Aside from that, it is poorly selective, since any impurities contained in the gas having an acidic character will affect pH.

4.1 CHARACTERISTICS OF REVERSE HEAD-SPACE ANALYSIS

Janák, Novák et al.[4,5] and Dravnieks and Krotoszynski.[6,7] The equilibrium concentration in volatile liquids was suggested and developed for the determination of organic trace impurities in gases by Ioffe and Vitenberg with Borisov and Kuznetsov.[8-12]

There are considerable differences between these two types of equilibrium concentration. The direct introduction of the solution of a nonvolatile solvent into a gas chromatograph is undesirable. The accumulation of a nonvolatile liquid in a chromatographic column can lead to a considerable increase in the retention parameters and a sharp drop in the separation efficiency, as it is difficult to obtain an instantaneous extraction of the volatile substances from a nonvolatile solvent. Therefore, with nonvolatile liquids the equilibrium concentration is accomplished by frontal saturation of a thin liquid film applied to a solid support[3-5,7] (a kind of chromatographic packing placed into a tube with an internal diameter of several millimeters) or applied directly to the walls of a tube.[6] Subsequently, the concentrate in the nonvolatile liquid is subjected to thermal desorption into the carrier-gas flow, while the total quantity (not the concentration) of the volatile components absorbed by the concentrator is determined once.

Equilibrium concentration in volatile liquids is achieved by purging a volume of the homogeneous liquid with the gas sample. The basic difference between this and the frontal version used for nonvolatile liquids lies in the saturation process in which the concentration of an impurity in the liquid increases uniformly in the entire volume. Also, the concentrated solution in a volatile solvent can be introduced directly into a chromatograph, repeating the analysis the necessary number of times and then establishing the amount of the impurities present by the common methods of quantitative analysis.

The various types of equilibrium concentration in nonvolatile and volatile liquids are based on different principles and are carried out in different ways. They are also characterized by various possibilities to increase the sensitivity and selectivity of the analysis.

A significant advantage of equilibrium concentration is the possibility of determining unstable impurities, even when they undergo chemical transformation during sample handling. In such cases, any method based upon total substance extraction in a concentration process can lead to large error. Gas samples containing unstable substances can be collected without distortion if the amount of impurity retarded by the concentrator is compensated by adequate portions of gas. Apparently, in these cases samples must be

170 EQUILIBRIUM CONCENTRATION OF IMPURITIES OF GASES

collected under such conditions that the rate of introduction of the substance into the concentrator exceeds the rate of its dissolution there.

4.2 EQUILIBRIUM CONCENTRATION IN NONVOLATILE LIQUIDS

The equilibrium concentration process outwardly differs somewhat from the methods of total substance collection from gases. A principal difference in these methods lies in the fact that total collection by purging the concentrator with the investigated gas can be accomplished before its passage through a layer of sorbent, that is, before the front of the substance being determined reaches the exit of the concentrator column (Fig. 4.1a). On the other hand, the equilibrium concentration process is accomplished by passing the gas through an absorber until the entire front of the analyzed

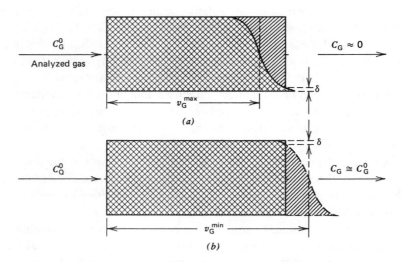

Figure 4.1. Frontal saturation of a sorbent layer using the method of (a) total collection, and (b) equilibrium concentration. C_G^0, concentration of the impurity in the gas being investigated; C_G, concentration of the impurity in the gas eluting from the concentrator column; δ, maximum permitted concentration of a substance eluting from the column; v_G^{max} and v_G^{min}, gas volume conducted through the concentrator which is maximal for the total collection method and minimal for the equilibrium concentration method.

4.2 EQUILIBRIUM CONCENTRATION IN NONVOLATILE LIQUIDS

impurities passes a layer of sorbent and the impurity concentrations in the gas flow at the entrance and exit of the concentrator become equal (Fig. 4.1b). In the total collection mode the initial concentration of the impurity is calculated from the volume of the purging gas which must be measured. In the total collection mode this purging volume must not be allowed to exceed a certain maximum volume (v_G^{max}), at which the passage of a substance will reach a maximum allowable value δ. On the other hand, in the case of equilibrium concentration it is important to purge only a minimum volume of gas, securing the achievement of the equilibrium concentration of an impurity in the entire volume of the packing material. Under these conditions, the volume in the tube of absorbing liquid V_L will contain $C_G^0 K V_L$ of the impurity being concentrated where C_G^0 is the impurity concentration in the gas investigated. In addition to this, the gas that fills the free volume V_G of the absorption tube will also contain $C_G^0 V_G$ of the impurity. Thus, the total quantity in the concentrate will be

$$q = C_G^0(KV_L + V_G) \qquad (4.1)$$

Here, K represents the distribution coefficient at the collecting temperature; V_L is the liquid volume in the concentrator column. If the column is filled with an adsorbent, V_L becomes the surface area of the adsorbent, and K the value of the adsorption coefficient; V_G now represents the gas-space volume of the concentrator, including the gas paths to the evaporator of the gas chromatograph.

If K is known, Equation (4.1) allows the calculation of the initial concentration of the impurity in a gas. The value of q is obtained after the elution of the substance into the gas chromatograph from the peak area, usually by the method of absolute calibration, while V_L and V_G are being measured during the preparation of the concentrator.

The absolute value of K can be found by one of the methods described earlier (see Section 1.2). However, in this case it should be used extremely carefully. The problem lies in the fact that the quantity of an impurity collected on a packing is determined not only by the effect of dissolution in the liquid phase, but also by the adsorption on the surface of the liquid or solid phase. Therefore, instead of K for the calculation of C_G^0, it is better to use parameters describing the retention of the analyzed substance in a chromatographic column having the same dimensions and packing as the concentrator. For instance, V_R^0, the retention volume, automatically considers all effects (both dissolution and adsorption) that determine the quantity of the collected substance. Furthermore, it is related to the distribution coefficient by the relationship

$$V_R^0 = KV_L + V_M^0 \qquad (4.2)$$

where V_M^0 is the retention volume of a nonadsorbed substance, which, together with (4.1) gives

$$C_G^0 = \frac{q}{V_R^0} \qquad (4.3)$$

For the purposes of calculation it is more convenient to use the specific retention volume (V_g)

$$V_g = \frac{V_R^0 - V_M^0}{w_L} \left(\frac{273.2}{T}\right) \qquad (4.4)$$

where w_L is the mass of the liquid stationary phase in the concentrator column. The derived equation for C_G^0 acquires the form

$$C_G^0 = \frac{q}{V_g w_L (T/273) + V_M^0} \qquad (4.5)$$

The use of V_g is convenient, because its value can be determined more accurately on a column of convenient length considerably different from the dimensions of the concentrator column.

The minimum gas volume needed to achieve equilibrium concentration along the entire length of the concentrator is conveniently expressed by the values of chromatographic retention obtained in test experiments and used repeatedly for the given concentrator column. In elution chromatography, v_G^{min} is the gas volume required for the elution of the total chromatographic band (Fig. 4.2) where ΔV_R is the gas volume necessary for the elution of half of the chromatographic band:

$$v_G^{min} = V_R^0 + \Delta V_R \qquad (4.6)$$

The values of V_R^0 and ΔV_R can be measured on the chromatogram in a test experiment. If it is difficult to carry out such measurement using short concentrators, then the value of the minimum gas volume is determined from the equation

$$v_G^{min} = V_R^0 \left[1 + \left(\frac{5.54 C u_0}{L}\right)^{1/2}\right] \qquad (4.6a)$$

where C is a constant characterizing the process of mass transfer, u_0 is the

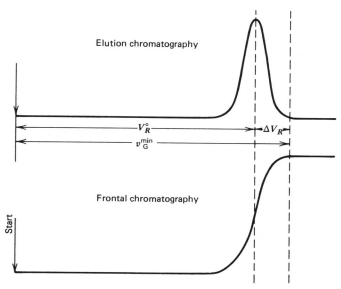

Figure 4.2. Minimum gas volume necessary to achieve equilibrium concentration of the impurity being collected.

linear carrier gas velocity, and L is the length of the packing layer in the concentrator. For instance, Novák et al.[3] used a concentrator 4.5 cm in length with 5 mm i.d., containing 0.3 g of packing (30% silicone elastomer E-301 on Celite), with a saturation rate of 300 ml/min and a linear velocity of 8.1 cm/sec. Substituting these values into equation (4.6a) we obtain that

$$(5.54 \ Cu_0/L)^{1/2} = 0.936$$

This means that v_G^{min}, the minimum gas volume which must pass through the concentrator is about twice of the retention volume V_R^0.

The sensitivity of the discussed method is determined by the relationship

$$S = A_L/C_G^0 \tag{4.7}$$

($A_L = fq$ is the peak area in the chromatogram that is equivalent to the mass of the eluted substance from the concentrator column $q = C_G^0 V_R^0$) and can be expressed in terms of the distribution (or adsorption) coefficient or the retention parameters:

$$S = f(KV_L + V_G) = fV_R^0 \tag{4.8}$$

Compared to the direct introduction of the investigated gas into the gas chromatograph the degree of sensitivity increase can be expressed as

$$\alpha = \frac{A_L}{A_G} = \frac{V_R^0}{v_g} \qquad (4.9)$$

where $A_G = fC_G^0 v_g$ is the peak area obtained in the direct introduction of the investigated gas of volume v_g into the gas chromatograph.

In order to increase the sensitivity of the analysis, parameters related to the concentrator column must be selected such that V_R^0 will be large. This is achieved by selecting an absorbing liquid with a large value of K and by increasing the volume of the packing (liquid phase). However, the increase in these two factors leads to an increased elution time for the substance from the concentrator column during thermal desorption. This decreases the efficiency of the chromatographic separation.

Generally speaking, a partial introduction of the desorbed substance into the separation column can be achieved by purging the concentrator column with a small volume of carrier gas equal to the volume of the gas phase in the concentrator (V_M). With this condition, only that part of the collected substance passes into the chromatographic column that is transformed into the gas phase at the desorption temperature. In this case the mass of the substance introduced into the chromatograph is determined by the difference between the collected quantity and that remaining in the concentrator packing.

$$V_M^0 C_G' = KV_L C_G^0 + V_M^0 C_G^0 - K' V_L C_G' \qquad (4.10)$$

where the primes indicate values at the desorption temperature.

Expressing the values of the distribution coefficient in this equation through the corresponding retention volumes, one obtains

$$C_G^0 = C_G' \frac{(V_R^0)'}{V_R^0} \qquad (4.11)$$

This allows the calculation of the initial concentration using the retention volumes of a substance by the concentrator column at the collecting and desorption temperatures (the value C_G' can be obtained by the method of absolute calibration).

Nevertheless, partial elution of the substance from a concentrator into the chromatographic column can lead to a decrease in the sensitivity of the analysis. Also, there are possible errors due to entrapment of small quantities of substances from a liquid by the purging gas (especially with long

4.2 EQUILIBRIUM CONCENTRATION IN NONVOLATILE LIQUIDS

concentrators). Therefore, the most correct determinations of C_G^0 are obtained upon the complete removal of the substance from the concentrator in the process of thermal desorption.

The time required for the removal of the substance (t^{min}) depends upon the desorption temperature, the purge rate of the carrier gas, and the volume of the packing material in the concentrator. The desorption time is usually selected experimentally when the peak area ceases to increase on the chromatogram. It is also possible to evaluate t^{min} (or a necessary volume of the carrier gas) from the retention values of the substance from the concentrator column at the collecting and desorption temperatures. If the purge rate of the carrier gas in the chromatographic column, F_c, which is limited by the conditions of the chromatographic analysis, is known, then

$$t^{min} = \frac{v_G^{min}(V_R^0)'}{F_c V_R^0} \qquad (4.12)$$

It is possible to increase the mass of the packing in the concentrator column without decreasing the separation efficiency if the collected impurities are not eluted directly into the chromatographic column but instead of this, are first collected in a capillary tube cooled with liquid nitrogen.[6,7] Upon rapid heating of the capillary in the carrier gas flow, an instantaneous sample introduction occurs. Total analysis time is lengthened, but with the intermediate stage the efficiency of the chromatographic column is considerably increased. Therefore, the sensitivity of the method is increased not only due to the increase in the packing mass, but also due to the compact introduction of the sample into the chromatographic column.

Novák, Vašák, and Janák,[3] who first used the equilibrium concentration in nonvolatile liquids, demonstrated the method using experimental mixtures containing the simplest aromatic hydrocarbons, acetone, and methanol in concentration ranges of 0.3–0.003 mg/liter. Direct introduction of the investigated gas into the chromatograph has been used for comparison. A glass tube of 5 mm i.d., 45 mm long, containing 0.3 g of the packing material was used as the concentrator column. For hydrocarbons, 30% silicone elastomer E-301 on granular Celite 545 0.125–0.25 mm was used as the packing, while for the polar substances, 30% polyethylene glycol 400 on the same solid support was used. Saturation of the packing with the investigated gas was accomplished at a rate of 300 ml/min at room temperature (22–25°C). Desorption occurred at 200°C for nonpolar and 150°C for polar substances.

The results of the analysis of the mixture, which contained the above-

Table 4.1 Results of the Analysis of Standard Mixtures of Organic Impurities in Air by Means of Direct Introduction of 1 ml of Gas into the Chromatograph and with a Preliminary Equilibrium Accumulation in the Column Concentrator[a]

Series	Substance	Concentrator Temperature, °C	V_g, ml/g	Concentration, mg/m³		
				Given	Direct Introduction of Gas	Equilibrium Concentration
1	Benzene	25	320	329	273	287
2	Benzene	24	335	110	124	114
	Toluene		930	38	37	37
	p-Xylene		2730	13	6	7
3	Benzene	22	360	36	35	37
	Toluene		1025	13	—	12
	p-Xylene		3050	4	—	3
4	Acetone	25	273	—	251	254
	Methanol		780	—	19	21
	Toluene		1030	—	38	38

[a]Conditions of concentration and gas-chromatographic analysis are given in the captions of Figures 4.3 and 4.4.

mentioned substances in various concentrations, are given in Table 4.1 and indicate good agreement with the direct-introduction method. The deviation does not exceed 10%. Table 4.1 also lists the established values of V_g, applied to 1 g of the liquid phase, characterizing the sensitivity and selectivity of the method. These data (considering a packing mass of 0.3 g) under the mentioned analytical conditions in conjunction with Equation (4.9) were used to calculate the sensitivity increase achieved by collecting the substance in the concentrator (at 25°C): benzene, 10; toluene, 30; p-xylene, 90; acetone, 9; and methanol 25 times (the value of v_g, in this case the maximum volume of the gas sample directly introduced into the chromatographic column, is assumed to be 10 ml). The gain in sensitivity of the analysis is well illustrated by the chromatograms given in Figures 4.3 and 4.4.[3] From these chromatograms it follows that the peaks of the "heavy" components (with large V_R^0) increase to a greater degree than those of the

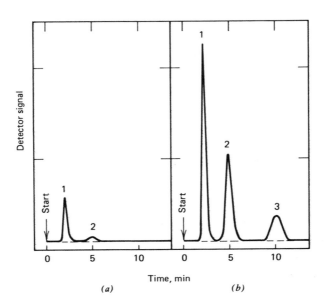

Figure 4.3. Chromatograms of *1* benzene, *2* toluene, and *3* p-xylene present as impurities in air.[3] (*a*) Direct introduction of 10 ml of air sample into the gas chromatograph; (*b*) Equilibrium concentration in a tube containing 0.3 g of a column packing consisting of 30% E-301 silicone on Celite, at a temperature of 24°C. Conditions of the gas-chromatographic analysis: Column: 170 cm × 6 mm i.d., containing 25% Apiezon L on Celite 545 (0.125–0.25 mm fraction). Column temperature: 100°C. Carrier gas (nitrogen) flow rate: 81.6 ml/min. Flame-ionization detector. Both chromatograms were obtained using the same detector attenuation.

178 EQUILIBRIUM CONCENTRATION OF IMPURITIES OF GASES

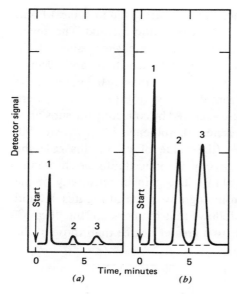

Figure 4.4. Chromatograms of (1) acetone, (2) methanol, and (3) toluene present as impurities in air.[3] (a) Direct introduction of 10 ml of air sample into the gas chromatograph; (b) Equilibrium concentration in a tube containing 0.3 g of a column packing consisting of 30% polyethylene glycol 400 on Celite, at a temperature of 24°C. Conditions of the gas-chromatographic analysis: Column: 85 cm × 6 mm i.d., containing 30% polyethylene glycol 400 on Celite 545 (0.125–0.25 mm fraction). Column temperature: 70°C. Carrier gas (nitrogen) flow rate: 50 ml/min. Flame-ionization detector. Both chromatograms were obtained using the same detector attenuation.

light components. From the chromatogram of the illuminating gas (Fig. 4.5), it is evident that upon direct introduction of 10 ml of the gas, the trace components are not recorded on the chromatogram. The possibility of the selective increase in sensitivity of the analysis is well indicated by the samples of gasoline vapors in air, Figure 4.6.

Equilibrium concentration on chromatographic sorbents has been used for the analysis of atmospheric impurities. For example, benzene, chlorobenzene and nitrobenzene in air,[4] and Halothane (2-chloro-2-bromo-1,1,1-trifluoroethane, an anesthetic) have been determined in the air of surgical rooms.[5] The method allows the determination of aromatic compounds in concentrations as low as several milligrams per cubic meter with an error ~25% (relative units). Halothane was determined using a packing material composed of 25% Apiezon K (0.45 g in a layer length of 5 cm) on Sterchamol and of Porapak P and Q (0.1 g each at the same layer length). The best results were obtained using Porapak Q. The degree of enrichment in comparison with direct introduction of the investigated air is greater than two orders of magnitude, and the minimum concentration determined using the flame-ionization detector is 0.01 ppm. (Halothane is usually present in the order of 10–50 ppm in surgical rooms.)

4.2 EQUILIBRIUM CONCENTRATION IN NONVOLATILE LIQUIDS

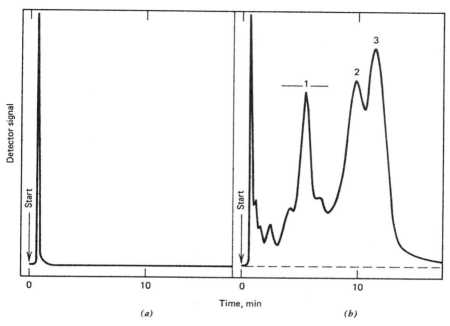

Figure 4.5. Chromatograms of trace components in illuminating gas.[3] (*a*) Direct introduction of 10 ml of air sample into the gas chromatograph. (*b*) Equilibrium concentration in a tube containing 0.3 g of a column packing consisting of 30% E-301 silicone on Celite, at a temperature of 24°C. Conditions of the gas-chromatographic analysis: Column: 150 cm × 6 mm i.d., containing 25% Apiezon L on Celite 545 (0.125–0.25 mm fraction). Column temperature: 150°C. Flame-ionization detector. Peaks: *1*, tri- and tetramethylbenzenes, *2*, indene, *3*, naphthalene.

When selecting the sorbents for equilibrium concentrations of atmospheric impurities, it is necessary to consider their hydrophilic properties. Water vapor contained in the air will be absorbed by the packing, considerably changing its sorption properties, which can result in large errors.[3] Therefore, to prevent the damaging effect of moisture in the equilibrium concentration of impurities in an air sample, it is necessary to use hydrophobic materials. The packing materials can include nonpolar liquid phases, but the best are polymers such as Porapak[5] and Tenax. Intermediate accumulation of the substance in a cryogenic concentrator followed by thermal desorption allowed the use of comparatively large spiral concentrator.[6,7] These were made of glass (25 cm long and 8 mm i.d.) with a thin film of nonvolatile liquid (Apiezon, diethylene glycol adipate) and were used to

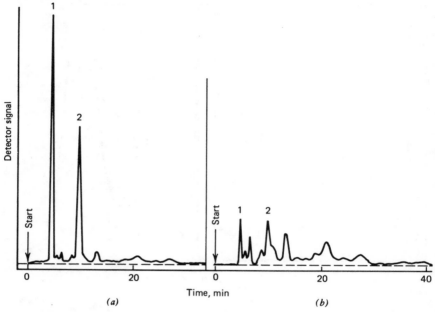

Figure 4.6. Chromatograms of gasoline vapors in air.[3] (*a*) Equilibrium concentration in a tube containing polyethylene glycol 400 (see caption of Fig. 4.4). (*b*) Equilibrium concentration in a tube containing E-301 silicone (see caption of Figure 4.3). Chromatographic conditions as in Figure 4.5. Column temperature: 60°C. Peaks: *1*, benzene, *2*, toluene.

determine atmospheric organic impurities at the ppb level and lower. Such high sensitivity of the analysis is often considerably lower than the sensitivity threshold of smell. This fact permitted to use the method for the identification of various objects in a mixture of volatile substances being released into the air.

4.3 EQUILIBRIUM CONCENTRATION IN VOLATILE LIQUIDS

The use of volatile solvents for equilibrium concentration of gas impurities[8–14] eliminates certain complications related to the necessity of thermal elution of the accumulated impurities from nonvolatile liquids or adsorbents.

The method is accomplished by purging a small volume of a pure volatile

4.3 EQUILIBRIUM CONCENTRATION IN VOLATILE LIQUIDS

liquid contained in a special saturator or vial (from multiliter volumes to several milliliters) with the investigated gas. The solution obtained is analyzed directly by introducing the usual aliquot (several microliters) into the evaporation chamber of the chromatograph. In other words, only the mass present in an aliquot of the solution of the collected impurity is introduced, not the entire mass, and repeated analytical determination of the concentrate is possible.

The basic difference between this and the method described in the previous section lies in the determination of the concentration of the collected substance as opposed to its total quantity. In addition, it is possible to select a suitable solvent from a large number of compounds. Also, the exclusion of thermal desorption allows the analysis of unstable substances.

The dynamics of impurity accumulation in a volatile liquid during the gas-purging process must be discussed. The gas contains an impurity in the concentration C_G^0 with a distribution coefficient K. Basically, this process is similar to continuous gas extraction of the volatile solutions discussed earlier (Section 1.4). The difference lies in the mass balance during the fundamental process of passing microbubbles of the gas through a solution. The mass of the impurity entering the liquid is $C_G^0 f_s dv_G$. Here $f_s dv_G$ is the volume of the bubble upon introduction into the liquid of pressure p, not including the liquid vapors of vapor pressure p_L, and therefore is a fraction f_s of the volume of this bubble dv_G upon leaving the liquid.

$$f_s = \frac{p - p_L}{p} \tag{4.13}$$

Using the same assumptions as in the discussion of continuous gas extraction (see Chapter 1, Section 1.4), the equation of mass balance can be written as

$$C_L V_L + C_G^0 f_s dv_G = (C_L + dC_L)(V_L + dV_L) + \frac{C_L + dC_L}{K} dv_G \tag{4.14}$$

Upon substituting the values V_L and dV_L from (1.61) and (1.61a) into this equation, separating the variables, and integrating the volume of the bubbled gas over the limits of 0 to v_G and over the concentration limits from 0 to C_L, one obtains

$$C_L = KC_G^0 \frac{f_s}{1 - FK} \left[1 - \left(1 + F\frac{v_G}{V_L^0}\right)^{(1-FK)/FK}\right] \tag{4.15}$$

Formula (4.15) connects the impurity concentration in a solution after purging with a certain volume of gas to the content of the trace impurity in it, its distribution coefficient, the initial volume of the liquid, and volatility F. Analysis of Equation (4.15)[9,10] leads to important conclusions concerning possible variations of the method of equilibrium concentration in volatile liquids. It also points out limitations of the method related to the properties of the absorbing liquids and the impurities to be concentrated.

1. In the case of a nonvolatile liquid, $p_L = 0$, and therefore $f_s = 1$, but $F = 0$. When $F \to 0$, Equation (4.15) becomes

$$C_L = KC_G^0 \left[1 - \exp\left(-\frac{v_G}{KV_L^0}\right) \right] \quad (4.16)$$

At the permitted error δ (in fractions) and a sufficiently large volume of purging gas

$$V_G > - KV_L^0 \ln \delta$$

For calculational purposes, one can use the simplest relationship, $C_L = KC_G^0$.

2. The last drop of a volatile liquid will evaporate when $v_G = V_L^0/F$. For $FK < 1$, the maximum impurity concentration in the liquid at that moment will be

$$C_L^{\lim} = KC_G^0 \frac{f_s}{1 - FK} \quad (4.17)$$

However, for the analysis it is necessary to have a certain final quantity of liquid ($V_L^0 - Fv_G$). It is important to clarify the character of the approximation C_L regarding the maximum value, Equation (4.17), in proportion to the bubbling of the investigated gas and evaporation of the absorbing liquid. The shape of the curves of various impurity concentrations in the absorbing liquid [Equation (4.15)] is considerably different depending on the volatility of the liquid (F) and the distribution coefficients of the impurities.

(a) At $FK < 0.5$, the curves $C_L(v_G)$ are convex relative to the positive direction of the axis C_L (Fig. 4.7). The derivative $\partial C_L/\partial v_G$ at $v_G = V_L^0/F$ goes to zero, and the concentration of impurities is near the limit (4.17) even before total evaporation of the liquid. Therefore, in principle the analysis can be accomplished without determining the volumes of the gas and liquid using formula (4.17).

(b) If $FK = 0.5$, then Equation (4.15) describes the straight line separating the convex and concave curves.

4.3 EQUILIBRIUM CONCENTRATION IN VOLATILE LIQUIDS   183

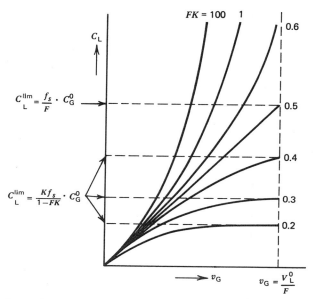

Figure 4.7. Curves showing the impurity concentration in the absorbing liquids, for various values of FK.

(c) If $FK > 0.5$, then the curves $C_L(v_G)$ are concave, and the derivative $\partial C_L/\partial v_G$ at the moment of total evaporation of the liquid goes to infinity. In this case, the final limiting value of impurity concentration corresponding to (4.17) will be achieved under conditions of a sharp rise in the curve $C_L(v_G)$. This occurs in the area directly preceding the total evaporation of the liquid only for values $0.5 < FK < 1$.

Analysis of the limit of Equation (4.15) as $FK \to 1$ gives the expression

$$C_L = C_G^0 K f_s \ln\left(\frac{V_L^0}{V_L^0 - F v_G}\right) \qquad (4.18)$$

which describes the concave line, separating the curves, and has a finite limit at $v_G \to (V_L^0/F)$ and an infinite limit at $v_G \to (V_L^0/F)$.

(d) If $FK \geq 1$, that is, for more volatile absorbing liquids with large distribution coefficients, the concentration of impurities in a liquid phase at $v_G \to (V_L^0/F)$ will increase without limit. In this case volatile liquids cannot be used for equilibrium concentration, and Equation (4.15) is not valid.

Thus, the conditions for the application of the equilibrium concentration method using volatile liquids involve the extent of the volatility of the absorbing liquid (at the saturation temperature). This value must be less than half the value of the volatility of the substance being determined, as characterized by the value $1/K$,* that is, $F < \frac{1}{2} K$, or

$$FK < 0.5 \tag{4.19}$$

If the permissible error of analysis is $\pm 100\delta\%$, in other words, at

$$\left(\frac{V_L^0 - Fv_G}{V_L^0}\right)^{(1-FK)/FK} < \delta \tag{4.20}$$

one can disregard the factor depending upon the volume of the gas in Equation (4.15) and v_G does not have to be measured. It is advantageous for the analysis that the fraction ∇ taken from the volume of the absorbing liquid should follow the criterion of

$$\nabla^{(1-FK)/FK} < \delta$$

or

$$FK < \frac{\ln \nabla}{\ln \nabla + \ln \delta} \tag{4.21}$$

Condition (4.20) determines the possibility of calculating the results of the analysis using formula (4.17) without measuring the gas volume. However, Equation (4.21) can be treated differently. When F and K ($FK < 0.5$) are given, one must bubble a gas volume of not less than

$$v_G^{\min} = \frac{V_L^0}{F} (1 - \delta^{FK/(1-FK)}) \tag{4.22}$$

in order that the concentration of the trace impurity in the liquid does not differ from the limiting concentration described by Equation (4.17) by more than $100\delta\%$ (Fig. 4.8).

Using the basic ratio ($K = C_L/C_G$) for volatile liquids assumes the fulfillment of additional limitations, consisting in the requirement

$$\left| \frac{f_s}{1 - FK} - 1 \right| < \delta$$

---

*The value $1/K$ indicates how many times the impurity concentration in the gas exceeds its equilibrium concentration in the liquid. Thus, it serves as a measure of the volatility of an impurity in the solution.

4.3 EQUILIBRIUM CONCENTRATION IN VOLATILE LIQUIDS

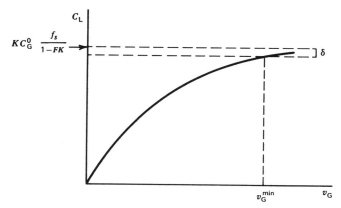

Figure 4.8. Minimum gas volume necessary to achieve a practically constant concentration of the collected impurity in the solution.

or

$$\frac{p_L - p\delta}{p - p\delta} < FK < \frac{p_L + p\delta}{p + p\delta} \qquad (4.23)$$

It is obvious that if $\delta > p_L/p$, then the left-hand inequality is fulfilled at any value of FK. In most cases, the permissible error in the analysis of determined impurities (δ) exceeds the ratio of the partial pressure of the liquid used for concentration to the external pressure. Therefore, the limitation to be considered is only the right-hand inequality in (4.23).

In such a way the simplest formula of the distribution law can be used in analytical calculations using the method of equilibrium concentration for substances with intermediate values of K. The upper limit of these values is determined by the volatility of the liquid and the permissible error δ.

The maximum value of impurity concentration achieved in a liquid C_L depends on the value ∇F (Fig. 4.9). This value is numerically equal to the volume increase of the gas phase during the evaporation of a single volume of liquid:

$$\nabla F = \frac{1 - f_s}{F}$$

Then Equation (4.17) can be written as

$$C_L^{\lim} = KC_G^0 \frac{1 - F \cdot \nabla F}{1 - FK} \qquad (4.24)$$

186 EQUILIBRIUM CONCENTRATION OF IMPURITIES OF GASES

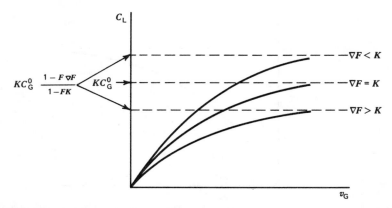

Figure 4.9. Dependence of the limiting concentration in a volatile solvent on the relationship between the values of ∇F and K.

From Equation (4.24) it follows that the relationship of the parameters ∇F and K determines the maximum obtainable values of C_L. At $\nabla F > K$, the impurity equilibrium concentration in the volatile solvent is lower than in the nonvolatile solvent possessing the same value of K. If $\nabla F < K$, the concentration C_L obtained by the saturation of the volatile solvent exceeds that derived from (1.3). In the particular case when $F = K$, formulas (1.3) and (4.24) will give identical results.

The sensitivity of the equilibrium concentration in volatile liquids, similar to (4.7), is expressed by the equation

$$S = f v_L K \frac{f_s}{1 - FK} \tag{4.25}$$

The basic factor affecting the sensitivity of the analysis is the value of K, since the factor $f_s/(1 - FK)$ in cases suitable for equilibrium concentration in volatile liquids ($FK < 0.5$) differs only slightly from 1. When determining heteroorganic substances using selective detectors, f can give a considerable input into the value of S.

The increase in the sensitivity of the analysis compared to the direct introduction of the investigated gas into the chromatograph is expressed by the relationship

$$\alpha = \frac{v_L}{v_g} K \frac{f_s}{1 - FK} = \frac{K f_s}{10^3 (1 - FK)} \tag{4.26}$$

It follows that increased sensitivity is achieved in the determination of

4.3 EQUILIBRIUM CONCENTRATION IN VOLATILE LIQUIDS

substances with K larger than 10^3. However, it is not advisable to use this method for substances with $K > 10^4$ since satisfying the condition $FK < 0.5$ requires the application of relatively nonvolatile solvents as well as fairly large volumes of purging gas. When achieving equilibrium the latter considerably increases the time of sample collection (several hours). Therefore, the possibility of lowering the threshold of determination by the direct introduction of the concentrate into the chromatograph only slightly exceeds one order of magnitude. Thus, variations that concentrate the impurities on the chromatographic packing are preferred (as a result of introducing only a fraction of the total mass of the impurity absorbed in the solvent).

The correct selection of the volatile liquid used for concentration is the deciding factor for the successful completion of the analysis using the equilibrium concentration method. The basic criterion by which one must be guided when selecting the solvent is the value of the distribution coefficient of the substances being determined in the solvent. To increase the sensitivity of the analysis the most useful values of K are in the range of 10^3–10^4. The solvent, too, must satisfy certain requirements as a result of the theory of equilibrium concentration in volatile liquids. According to these requirements the possibility of achieving a practically constant concentration of an impurity in a solution is determined by the constancy of the distribution coefficient in the investigated ranges of concentrations and the condition $FK < 0.5$. It is beneficial if the peak of the solvent selected is not recorded on the chromatogram. This can be achieved by selectively retarding the solvent in a precolumn or by applying selective detectors with low sensitivity toward the solvent. If these steps are impossible, the conditions of separation should be chosen such that the solvent peak does not overlap the peaks of the impurities being determined. In selection of the absorbing liquid, both the availability and possibility of simple purification from interfering impurities become important considerations.

Table 4.2 lists the properties of certain liquids used for the determination of trace impurities in gases by the equilibrium concentration method.[10-14] The table also includes the limiting values of the distribution coefficients of the impurities corresponding to the boundaries of the preferred application of the equilibrium concentration method according to condition (4.19). The limiting values of K are also given for the use of $C_G^0 = C_L^{lim}/K$ with an accuracy of ±10%. This accuracy is quite acceptable when working at the maximum acceptable levels of impurity concentrations in atmospheric samples.

As it is evident, water is an especially convenient absorbing liquid due to

Table 4.2 Volatility Characteristics of Some Liquids (at 20 and 30°C and Impurities for the Analysis by the

Absorbing Liquids	Boiling Point, °C	p_L, mm Hg		f_s	
		20°C	30°C	20°C	30°C
Water	100	17.5	31.8	0.977	0.958
Acetic acid	118	11.8	20.0	0.985	0.973
n-Butyl alcohol	118	4.4	9.3	0.994	0.988
Benzene	80	75.2	118.0	0.901	0.845

[a]At $\delta = 0.1$ (10%).

its relatively low values of F related to its low molecular mass. The limitations applicable to the properties of acetic acid are more pronounced. For example, for impurities with distribution coefficients exceeding 1560, the simplest formula of the distribution law creates errors larger than 10%, and one must use Equation (4.17).

The basic tenets of the theory of equilibrium concentration in volatile liquids have been investigated using samples of gaseous aromatic hydrocarbons absorbed by acetic acid and certain oxygen compounds absorbed by water.[11,12]

Equation (4.15), describing the accumulation of an impurity in a liquid purged by the contaminated gas, was derived for equilibrium conditions. Its correspondence to real processes is characterized by Figures 4.10 and 4.11. These graphs are plotted as fraction of nonevaporated liquid versus the relationship of the concentrations of an impurity in the liquid and the gas. The position of the curves for the equilibrium concentration in such coordinates depends only on the distribution coefficient of the impurities and the parameters characterizing the volatility of the liquids (Table 4.2). As it is evident from the results, good agreement is noted between experimental and theoretical calculations. The experimental points belonging to various experiments with different gas flow rates (190–420 ml/min), different impurity concentrations (0.7–14 mg/m³), and different volumes of absorbing liquids fall on the general curves corresponding to Equation (4.15) in the coordinate system of Figures 4.10 and 4.11.

The correlation with theory is also supported by the results of the determination of the maximum (equilibrium) concentrations. The deviation of results using Equation (4.17) and established values C_L^{lim} averages 3–

4.3 EQUILIBRIUM CONCENTRATION IN VOLATILE LIQUIDS

760 mm Hg Pressure) and Maximum Values of the Distribution Coefficients of Trace Equilibrium Concentration Method

$F \times 10^5$		∇F		K^{max}			
				Eq. (4.19)		Eq. (4.21)[a]	
20°C	30°C	20°C	30°C	20°C	30°C	20°C	30°C
1.73	3.07	1336	1378	28,900	16,300	7080	4210
6.73	11.06	229	238	7,430	4,520	1560	1040
2.20	4.54	263	269	22,700	11,000	4370	2245
36.6	56.2	270	277	1,370	890	490	410

6% (Table 4.3) and is within the limits of error for the conducted measurements.

When $FK > 0.5$ the maximum value of impurity concentration in solution at the moment of total evaporation of the solvent occurs under conditions creating a sharp rise in the curve and is observed when phenol is distributed between water and air (Fig. 4.12). In this case, the curve is convex

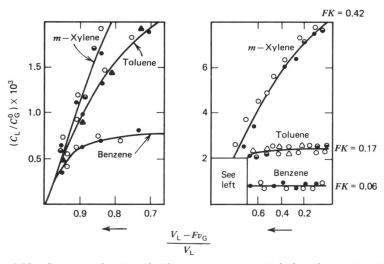

Figure 4.10. Saturation of acetic acid with air containing aromatic hydrocarbons as impurities at 25°C. The left-hand diagram illustrates on an enlarged scale the path of the curves before evaporation of one-third of the initial volume of acetic acid. The solid lines represent computed curves while the signs (●, ○, ◓ ▲ and △) represent measured data obtained under different experimental conditions. $C_G^\circ = 0.1$–14.3 mg/m^3; $V_L = 1.5$–5 ml; flow rate of gas purge: 190–422 ml/min.

Table 4.3 Values of C_L^{lim}, mg/liter, of Trace Impurities at 25°C, Obtained by the Equilibrium Concentration Method and Computed from Equation (4.17).

Solvent	V_L^0, ml	v_G, liter	Impurity	C_G^0 mg/m³	Exptl.	C_L^{lim}, mg/liter Using Eq. (4.17)	Deviation,%
Acetic acid	1.5–5	17–56	Benzene	1.20	0.94	0.92	2.1
				1.30	1.00	0.99	1.0
				2.04	1.53	1.56	2.0
				11.80	9.50	9.00	5.3
				18.00	9.35	9.90	5.9
				14.30	10.40	10.90	4.8
						Avg.	3.5
			Toluene	0.38	0.93	0.92	1.1
				0.70	1.57	1.69	7.7
				0.74	1.76	1.79	1.7
				1.02	2.22	2.45	10.4
				6.90	15.90	15.70	5.0
				8.80	20.00	21.30	6.5
						Avg.	5.4

Water						
1.5–2.5	Acetone	2.44	1.66	1.54		7.2
		3.24	1.84	2.05		11.4
		3.38	2.16	2.13		1.4
		3.94	2.64	2.52		4.6
		36.70	22.00	23.20		5.5
		37.80	23.00	23.90		3.9
					Avg.	5.6
10–20	Methyl ethyl ketone	1.29	0.59	0.56		5.1
		2.03	0.92	0.87		5.5
		2.05	0.81	0.88		8.6
		2.60	1.20	1.14		5.0
		27.40	10.90	11.75		7.8
		30.80	12.90	13.20		2.3
					Avg.	5.7

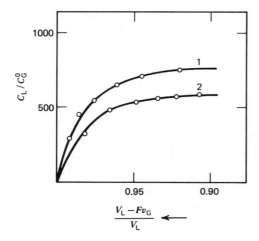

Figure 4.11. Saturation of 2 ml of water at 20.5°C with air containing 1 acetone (C_G^o = 50.2 mg/ m³) and 2 methyl ethyl ketone (C_G^o = 27.1 mg/m³) as impurities. Gas flow rate: 185 ml/min. Circles represent experimental results while the solid lines represent computed values.

to the abscissa and conforms with the experiment. For the determination of traces of phenol in air, such a curve is unfavorable; however, if the saturation conditions are changed in order to lower the volatility of the solvent so that $FK < 0.5$, then one can achieve an almost constant concentration of phenol in water. For example, if the air has a relative humidity of 50%, then F becomes equal to 6.55×10^{-5}, and $FK = 0.29$. At this point an area of stationary concentration appears on the saturation curve (Fig. 4.12b).

The equilibrium concentration in pure volatile liquids was first used[8] to determine trace impurities of aromatic hydrocarbons and carbonyl compounds in air. A suitable absorber for aromatic hydrocarbons happens to be acetic acid, and for carbonyl compounds, water.

These solvents have favorable values for the substances distribution coefficient values for the substances and are available in chromatographically pure form. They can be totally separated from the analyzed compounds by absorption in a precolumn containing potassium hydroxide.[15] Figure 4.13 presents chromatograms of the same dilute solution of aromatic hydrocarbons in acetic acid (a) without and (b) with the absorption of the solvent (acetic acid) in the precolumn. The time necessary for the chromatographic analysis of benzene, toluene, and m-xylene is less than 5 min. However, the total elution time of 1 μl of solvent under the conditions given in Figure 4.13 is approximately 2 hrs and thus, would completely overlap the peaks of interest. A similar effect is achieved for aqueous solutions. The only difference is that the packing of the precolumn does not completely absorb

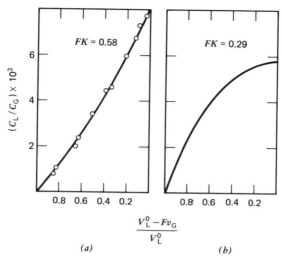

Figure 4.12. Saturation of 3 ml of water at 60°C with air containing phenol ($C_G^\circ = 23.6$ mg/m³) as the impurity. Gas flow rate: 198 ml/min. (a) Dry air, (b) air with a relative humidity of 50%. Circles represent experimental results while the solid lines represent computed values.

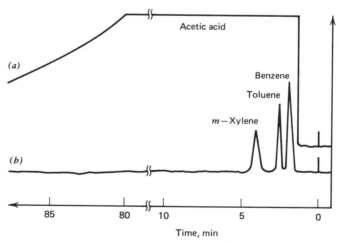

Figure 4.13. Chromatograms of the solution of aromatic hydrocarbons in acetic acid (concentration: 10^{-4}%) (a) without and (b) with absorption of the solvent on a precolumn.[15] Chromatographic conditions: Column: 100 cm × 4 mm i.d. containing 20% Apiezon L on Chromosorb W (0.1–0.2 mm fraction). Column temperature: 100°C. Carrier gas (argon) flow rate: 60 ml/min. Flame-ionization detector with a full-scale response of 2×10^{-12} A. Sample size: 1 µl. The precolumn (50 cm × 4 mm i.d.) contained 10% polyethylene glycol 600 on Spherochrome-1 (0.1–0.2 mm fraction) in the case (a); in the case (b) 20% potassium hydroxide was added to the packing of the precolumn.

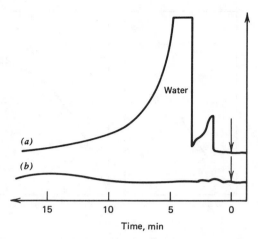

Figure 4.14. Chromatograms of distilled water.[15] Chromatographic conditions: Column: 100 cm × 4 mm i.d., containing 20% Tween 20 on Chromosorb W (0.1–0.2 mm fraction). Column temperature: 100°C. Carrier gas (argon) flow rate: 40 ml/min. Flame-ionization detector with a full-scale response of 1×10^{-11} A. Sample size: 1 μl. The precolumn (50 cm × 4 mm i.d.) contained 15% polyethylene glycol 600 on Spherochrome-1 (0.1–0.2 mm fraction) in the case (a); in the case (b) 20% potassium hydroxide was added to the packing of the precolumn. The broad peak from 10 minutes on corresponds to water.

the water, but selectively retains it, extending the elution of the water peak for a considerably long time (Fig. 4.14: see the broad peak starting at 10 minutes in Fig. 4.14b). If the concentration of water vapor in the carrier gas at the exit from the chromatographic column does not exceed 10^{-3}%, then the quality of the chromatogram is practically unaffected although it is recorded using a flame-ionization detector.

The analytical accuracies indicated in Table 4.4 are characteristic for the analysis of air containing vapors of various compounds. In all cases listed there $FK < 0.5$, and the concentration of the impurities in the solution is near the limit even before the total evaporation of the liquid. The analysis was accomplished without determining the volumes of the purging gas or the liquid. The results were calculated using formula (4.17), however, for the compounds with relatively small values of F and K (benzene in acetic acid, acetone, methyl ethyl ketone, and diethylamine in water), the calculations were completed using the simplest formula (1.3). The impurity concentrations in the equilibrium solutions were determined using a flame-ionization detector by the method of absolute calibration using the peak heights. The average deviation varied between 2.5 and 6.6%. For the given concentration ranges, this is considered completely acceptable.

4.3 EQUILIBRIUM CONCENTRATION IN VOLATILE LIQUIDS

Table 4.4 Determination of Trace Concentrations of Various Compounds Present in a Dry Air Sample*, Using Equilibrium Concentration in Acetic Acid (Hydrocarbons) and in Water (Carbonyl Compounds and Diethylamine)

Trace Impurity	Concentration Range, mg/m^3	Mean Relative Error of the Determination, [a]%
Benzene	0.6–14.3	4.2
Toluene	0.4– 8.0	2.5
m-Xylene	0.1– 2.9	3.6
Acetone	1.6–37.8	5.3
Methyl ethyl ketone	0.5–30.8	4.0
Methyl propyl ketone	1.0–11.5	6.6
Acetaldehyde	7.6–65.2	6.3
Diethylamine	0.9–61.2	5.1

[a]In comparison to the actual concentration.

*Samples were prepared using the diffusion method.

The principle of equilibrium concentration is used in the determination of impurities of the simplest sulfur-containing substances (methyl mercaptan, ethyl mercaptan, and dimethyl sulfide) in natural gas and air.[13] For the selective detection of sulfur-containing substances, a flame-photometric detector (FPD) was used. The solvent used for concentration was benzene. The selection of the absorber is limited both by the availability of benzene in a sufficiently pure state and the properties of the FPD, which is fairly insensitive to hydrocarbons. Besides that, under the conditions of the gas-chromatographic analysis (Fig. 4.15) benzene has a longer retention time than the sulfur-containing compounds and therefore does not inject additional errors into the quantitative interpretation of the chromatograms.

A fundamental advantage of the method is that it allows the analysis of unstable mercaptans that would be oxidized during the sampling process (concentration). The necessary condition for the analysis of unstable substances by equilibrium concentration is that the flow rate of the substance into the absorber exceeds the rate of its consumption in the solution. The accomplishment of this condition is indicated by the emergence of a saturation curve leading to a horizontal area. Figure 4.16 gives the saturation curves for the simplest sulfur compounds at various temperatures and concentrations using benzene as the solvent. In all cases the curves have a

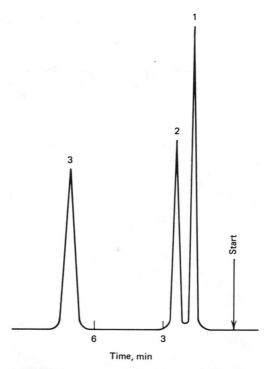

Figure 4.15. Chromatogram of the benzene solution of mercaptans. Chromatographic conditions: Column: 300 cm × 4 mm i.d., glass, containing 15% polyethylene glycol adipate on Celite C-22 80–100 mesh. Column temperature: 100°C. Carrier gas (helium) flow rate: 30 ml/min. Flame-photometric detector with a full-scale response of 1×10^{-9} A. Sample size: 2 µl. Peaks: *1*, methyl mercaptan; *2*, ethyl mercaptan; *3*, benzene.

constant concentration range after purging with a certain volume of gas. This is indicative of equilibration. The minimum volume of gas needed for this at $V_L^0 = 3$ ml does not exceed 8 liters.

Samples are collected at constant temperature (10–30°C) by bubbling the studied gas at a rate of 100–200 ml/min through 1–3 ml of benzene, which has been placed in a medical vial or a special saturator (Fig. 4.17). This container is equipped with an outlet fitted with a rubber septum for taking out samples of the solution and introducing them into the chromatograph with a microsyringe. The purging is increased simultaneously using several stainless steel capillary tubes with 0.4–0.6 mm i.d.

Upon completion of the purge, the solution is either immediately ana-

4.3 EQUILIBRIUM CONCENTRATION IN VOLATILE LIQUIDS

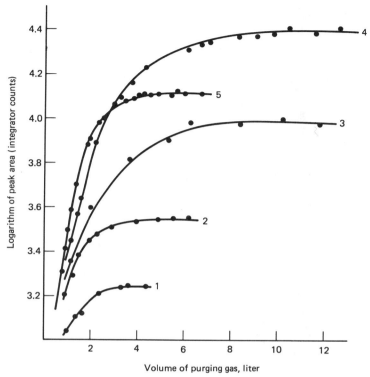

Figure 4.16. Saturation curves of sulfur-containing compounds present in air as impurities, using benzene as the solvent. *1, 2*, Methyl mercaptan; *3, 4*, ethyl mercaptan; *5*, dimethyl sulfide. Concentrations in air and temperatures: *1*, 0.13 mg/liter (15°C); *2*, 0.18 mg/liter (15°C); *3*, 0.19 mg/liter (15°C); *4*, 0.25 mg/liter (10°C); *5*, 0.16 mg/liter (20°C).

lyzed chromatographically for its content of the collected substances (usually by the method of absolute calibration) or it is sealed in a glass vial. Special attention is paid to the total absence of oxygen inside the vial, since it would quickly oxidize the mercaptans. Figure 4.18 illustrates that the amount of methyl and ethyl mercaptan in benzene solutions stored in contact with air decreases to one-fourth to one-sixth of the original value in 10 h. Glass vials with drawn ends (Fig. 4.19) are recommended to seal a solution without the gas phase. The vials are filled by pulling in the solution (using a rubber bulb), after which they are quickly sealed at both ends. If the benzene solutions are stored in vials in dark places, the concentration of the alkyl mercaptans will remain constant over several months.

198   EQUILIBRIUM CONCENTRATION OF IMPURITIES OF GASES

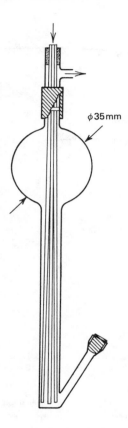

Figure 4.17. Concentrator for the accumulation of impurities of a gas in a solvent.

Concentrations of sulfur compounds in the studied gas are computed using Equation (4.17) with the numerical values of F and K given in Tables 4.2 and 4.5.

The test results of the determination of trace concentrations of sulfur compounds in air over the range of 0.05–0.3 mg/liter indicate that the error of the measurement does not exceed 10%. It was also demonstrated using dimethyl sulfide that the substitution of natural gas for air does not considerably affect the values of the distribution coefficients (Table 4.5). Therefore, the method of equilibrium concentration can also be used with natural gas as the sample. The limit of determination of the method depends mainly on the type of detectors used. Utilizing a flame photometric detector (FPD), the minimum detectable limit can be 1 mg/m^3 or even less.

The possibility of equilibrium concentration in volatile liquids for the

4.3 EQUILIBRIUM CONCENTRATION IN VOLATILE LIQUIDS

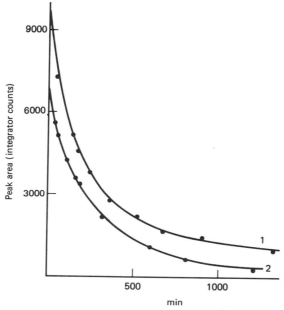

Figure 4.18. Concentration change of *1* methyl mercaptan and *2* ethyl mercaptan in benzene solution upon storage in contact with air.

selective accumulation of impurities and their separation from the accompanying substances is illustrated by the data in Table 4.6. As shown, the distribution coefficients of aromatic hydrocarbons between acetic acid and air vary from 690 to 5500, while the variation of carbonyl compounds is somewhat lower. Because of this, when acetic acid is employed as the concentration solvent, the solution will be enriched by the hydrocarbons in comparison to the carbonyl compounds. The effect is even greater for toluene and *m*-xylene than for benzene. The selective accumulation of aromatic hydrocarbons in acetic acid (proportional to the values of K) is illustrated in Figure 4.20a. For the chromatogram in Figure 4.20b the

Figure 4.19. Glass vial for the storage of solutions without a gas phase.

Table 4.5 Values of the Function $K \cdot f_s/(1 - FK)$ for Aromatic Hydrocarbons in Acetic Acid Solutions

Concentration of Acetic Acid, wt-%	Benzene			Toluene			m-Xylene		
	15°C	25°C	35°C	15°C	25°C	35°C	15°C	25°C	35°C
100	—	810	580	—	2670	1820	—	10390	7920
95	860	600	430	2470	1770	1180	6340	5210	3600
90	660	450	320	1740	1220	810	4030	3130	2130
85	500	350	240	1250	860	570	2690	2030	1360
80	390	270	180	900	610	400	1840	1350	910
75	300	200	140	650	440	290	1270	910	610

Table 4.6 Distribution Coefficients of the Simplest Aromatic Hydrocarbons in Liquid–Air Systems at 20°C

Hydrocarbon	Acetic Acid 100%	Acetic Acid 80%	Water	Aqueous Potassium Acetate (46%)
Benzene	895	313	4.8	0.7
Toluene	2479	709	4.6	0.6
m-Xylene	6369	1577	5.9	0.5

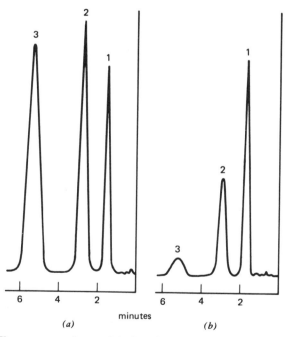

Figure 4.20. Chromatograms of aromatic hydrocarbons present in air, in concentrations of *1* benzene 13 mg/m³, *2* toluene 6.9 mg/m³, *3* m-xylene 2.5 mg/m³. (*a*) Direct analysis of the air sample using equilibrium concentration in acetic acid. (*b*) Absorption of the aromatic hydrocarbons on silica gel (total collection mode), extracting them with acetic acid, and analyzing the acetic acid solution. For chromatographic conditions, see Figure 4.13(*b*).

202　EQUILIBRIUM CONCENTRATION OF IMPURITIES OF GASES

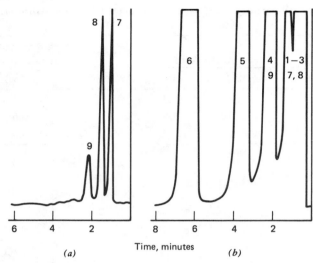

Figure 4.21. Chromatograms of impurities in air collected in two different modes: (a) dissolving in water (equilibrium concentration mode), (b) concentrating the impurities on silica gel (total collection mode) with subsequent displacement of the collected substances by acetic acid. The impurities present and their concentration in air: *1*, hexane (35 mg/m^3); *2*, heptane (22 mg/m^3); *3*, octane (15 mg/m^3); *4*, benzene (43 mg/m^3); *5*, toluene (15 mg/m^3); *6*, *m*-xylene (14 mg/m^3); *7*, acetone (3 mg/m^3); *8*, methyl ethyl ketone (4 mg/m^3); *9*, methyl propyl ketone (3 mg/m^3). For chromatographic conditions, see Figure 4.14(a).

aromatic hydrocarbons present in the air sample were first totally adsorbed on silica gel and then eluted by acetic acid. The separation of the undesired accompanying substances can be illustrated by the concentration of carbonyl compounds in water, in the presence of hydrocarbons. For water and air, the values of K for aldehydes and ketones exceed those of the hydrocarbons by more than two orders of magnitude. Therefore, during sample collection, an almost complete separation of the carbonyl compounds from the hydrocarbons which would hinder their gas chromatographic determination will occur. Thus, in the chromatogram of water saturated with air containing aromatic and aliphatic hydrocarbons and ketones (Fig. 4.21), the hydrocarbon peaks are practically absent, although the concentration of these compounds in the gas exceeds the concentration of the ketones by one order of magnitude. At the same time, if the aromatic compounds are concentrated on silica gel using the total collection mode, with subsequent displacement of the collected substances by acetic acid, the quantitative determination of the ketone content is impossible: these peaks are completely overlapped by the peaks of the accompanying hydrocarbons.

4.3 EQUILIBRIUM CONCENTRATION IN VOLATILE LIQUIDS

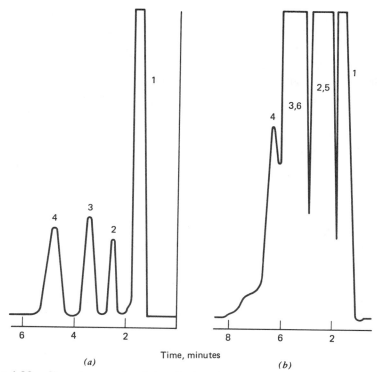

Figure 4.22. Chromatograms of a solution of hydrocarbons and carbonyl compounds in acetic acid. Chromatogram (a) was obtained by removing the carbonyl compounds on the precolumn containing lithium aluminum hydride while chromatogram (b) was obtained without the use of this precolumn. The precolumn for the removal of acetic acid was used in both cases. Chromatographic conditions: Analytical column: 100 cm × 4 mm i.d., containing 20% cyanoethylated pentaerythritol on Spherochrome-1 (0.125–0.20 mm fraction). Column temperature: 140°C. Carrier gas (argon) flow rate: 20 m/min. Flame-ionization detector, with a full-scale response of 1×10^{-12} A. Sample size: 2 μl. Precolumns: First (for the removal of acetic acid): 50 cm × 4 mm i.d., containing 10% polyethylene glycol 600 + 20% potassium hydroxide on Spherochrome-1 (0.20–0.35 mm fraction). Second (for the removal of the carbonyl compounds): 4 cm × 4 mm i.d., containing 2 g lithium aluminum hydride on glass wool. Compounds present (peaks) and their concentration in the solution: *1*, *n*-decane (24 mg/liter); *2*, benzene (0.65 mg/liter); *3*, toluene (0.85 mg/liter); *4*, *m*-xylene (1.1 mg/liter); *5*, ethanol (20.5 mg/liter) + isobutyraldehyde (12.4 mg/liter) + ethyl acetate (18.5 mg/liter) + acetone (12 mg/liter); *6*, methyl propyl ketone (19.3 mg/liter).

The given examples indicate the possibility of the separation of coexisting compounds at the concentration stage. Such separation is only possible if the values of the distribution coefficients sharply differ. If the components of complex mixtures have comparable K values, one must use highly efficient separation columns, including the use of capillary columns. Nevertheless, for serial analyses, especially when determining a narrow range of substances, it is more expedient to use selective chromatographic columns together with the methods of reaction gas chromatography. An example for this is the determination of aromatic hydrocarbons in a acetic acid solution containing comparable concentrations of paraffins and naphthenes and a series of oxygen-containing substances.[16] The separation of these compounds may be accomplished using an analytical column containing cyanoethylated pentaerythritol as the stationary phase which is preceded by a precolumn used for the retardation of acetic acid and possible amounts of other organic acids. The application of this superselective stationary phase allows the separation of most of the oxygen-containing substances from the aromatic, paraffinic, and naphthenic hydrocarbons.* However, on this column, lower alcohols, carbonyl compounds, and esters elute together with the aromatic hydrocarbons (Fig. 4.22). Therefore, they hamper and sometimes exclude the possibility of their determination. The best results in eliminating the oxygen-containing compounds can be obtained by using a second precolumn containing lithium aluminum hydride. As shown in Figure 4.22a, only the hydrocarbons are recorded in the chromatogram allowing the determination of the aromatic hydrocarbons with good accuracy ($\sim$5 rel. %).

4.4 EQUILIBRIUM CONCENTRATION IN VOLATILE LIQUIDS WITH VARYING VALUES OF THE DISTRIBUTION COEFFICIENTS

The theory of equilibrium concentration discussed in the previous section predicts the constancy of the values of K, F, and f_s when purging the liquid with the investigated gas. This condition is not fulfilled if the analyzed gas contains other impurities that affect the distribution coefficients of the components being determined. If the purging gas contains a very soluble impurity in large concentrations along with the impurity being determined, a gradual change occurs in the composition of the absorbing liquid. This

---

*Instead of cyanoethylated pentaerythritol, it is possible to use tris(cyanopropyl)amine which exhibits the same high selectivity in relation to the separation of paraffinic, naphthenic, and aromatic hydrocarbons.

4.4 VARYING VALUES OF THE DISTRIBUTION COEFFICIENTS

change results in a corresponding change in the distribution coefficient of the determined substance between the liquid and gas phases. The shape of the concentration curves will be determined through the relationship of the function $Kf_s/(1 - FK)$ to the composition of the absorbing liquid.[17] If a stationary concentration of the accompanying impurity is established in the solution after purging by a given volume of gas, then the accumulation process of the trace impurity being determined will be described by the curves of type *2* or *4* in Figure 4.23. The horizontal regions of these curves correspond to the equilibrium concentration of the trace impurity in the resulting solution. Depending on the character of the change in the distribution coefficient (increases or decreases of K according to the accumulation of the accompanying impurity), the concentration curves acquire the shape of curves *1* or *5* (Fig. 4.23) during a continuous increase of the accompanying impurity in the solution.

The general theory of the equilibrium concentration with varying distribution coefficients is not yet developed. One example studied is the accumulation of aromatic hydrocarbons in acetic acid from air.[17] The problem

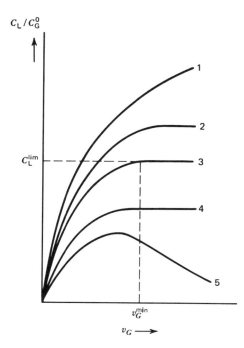

Figure 4.23. Influence of a compound present in the gas phase on the accumulation dynamics of a compound in solution (see explanation in the text).

lies in the dilution of acetic acid with water. Thus the distribution coefficients will decrease during the process of collecting samples when a hygroscopic solvent is being used in the analysis of a wet gas. Theoretical and experimental investigations[18] have indicated that curves having a maximum (type 5) can be obtained in such cases (Fig. 4.23). The position of the maximum determines the moment of reaching the equilibrium concentration of the trace impurity in the liquid characterized by the present value of K. This value depends upon the composition of the liquid phase. The descending area of the curve corresponds to the loss of the trace impurities from the solution due to a decrease in the distribution coefficient while diluting the acid with water.

Figure 4.24 illustrates the dependence of the concentration of aromatic hydrocarbons in a fluid on the volume of the purging gas with varying humidity. It is evident that the slope of the curves after the maximum increases with an increase in the humidity of the analyzed air. In this way, the highest concentration of the trace impurity in the liquid decreases and is achieved with a smaller volume of purging gas. The value of C_L/C_G^0 at any

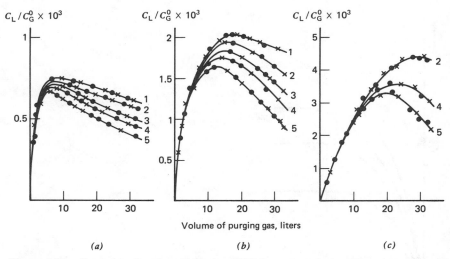

Figure 4.24. Saturation of acetic acid (3 ml at 25°C) by aromatic hydrocarbons from air, as a function of the humidity of the air. Compounds: (a) benzene; (b) toluene; (c) m-xylene. Humidities: 1, 0.95 mg/liter; 2, 1.25 mg/liter; 3, 6.8 mg/liter; 4, 9.4 mg/liter; 5, 23.0 mg/liter. Each curve includes experimental results with two different concentrations of the hydrocarbon (•, x), in the range of 1–40 mg/m³. The air purging rate was 200 ml/min.

4.4 VARYING VALUES OF THE DISTRIBUTION COEFFICIENTS

point in these regions of the curves at equilibrium is characteristic of the purging process. As such, the value of the ratio must be equal to the value of the function $Kf_s/(1 - FK)$ for the corresponding acetic acid concentration.

The relationship between the equilibrium concentration of the impurity in the solution and the descending region of the curve can be experimentally proven by comparing the calculated value of the function $Kf_s/(1 - FK)$ with the concentration in the solution. Figure 4.25 shows such plots for the accumulation of aromatic hydrocarbons in acetic acid from air of different humidity. As seen, the correlation between the descending regions of the concentration curves and the calculated values of the function $Kf_s/(1 - FK)$ is good.

Thus, the process of equilibrium concentration can be used to analyze trace impurities in gases even when the distribution coefficients change during sample collection. The method of determining aromatic hydrocarbons in humid air in this way consists of saturating the acetic acid with the air sample. Subsequently the concentration of the trace impurity in the liquid is determined by gas chromatographic analysis. Finally, the concentration of acetic acid is obtained by titration with a base or by gas chromatography. The concentration of the aromatic hydrocarbons in air can be calculated using Equation (4.17), with the numerical values of the function $Kf_s/(1 - FK)$ corresponding to the established acetic acid concentration.

For satisfactory result the absorbing liquid must be purged with a volume of a gas equal to or exceeding v_G^{min}, that is, the volume required to reach the descending region of the concentration curve. Accurate calculation of this volume is only possible if data concerning the humidity of the investigated gas are known. Such data are not always available. However, the value v_G^{min} decreases with an increase in the humidity of the gas (Fig. 4.23), and for practical applications it is satisfactory to use the value v_G^{min} obtained for dry air. Under these conditions one can always be sure that the concentration of aromatic hydrocarbons in the liquid corresponds to the values located on the descending regions of the concentration curves.

The described method for the determination of aromatic hydrocarbons was tested using standard mixtures of varying humidity with a known amount of benzene, toluene, and m-xylene.[17] The average deviation between the given and established concentration values in the range of 1–50 mg/m^3 is 5–7%. The limitation of determining these impurities in air mainly depends upon K, which is a function of both the concentration of the acid and the temperature of the sample. As the temperature and humidity of

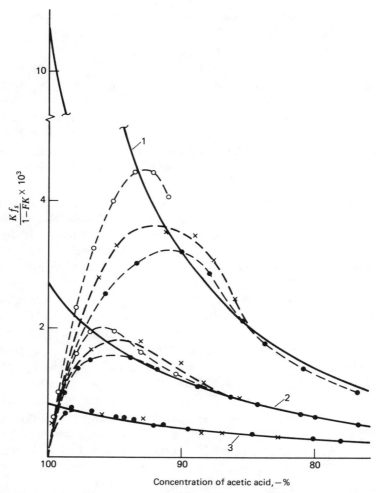

Figure 4.25. Dependence of the function $Kf_s/(1 - FK)$ (solid lines) for aromatic hydrocarbons on the concentration in acetic acid (broken lines). Hydrocarbons: *1*, *m*-xylene, *2*, toluene; *3*, benzene. Humidity of the air: ○ 1.25 mg/liter; x 9.4 mg/liter; ● 23 mg/liter. $V_L^\circ = 3$ ml; temperature: 25°C. The air purging rate was 200 ml/min.

the sample increase, K decreases. Correspondingly, the minimum detectable limit of aromatic hydrocarbons in the air increases. A gas sample at 25°C and 100% humidity has a limiting value of 0.3 mg/m^3 for benzene, 0.1 mg/m^3 for toluene and 0.5 mg/m^3 for *m*-xylene.

4.5 HEAD-SPACE ANALYSIS COMBINED WITH EQUILIBRIUM CONCENTRATION

The possibilities of equilibrium concentration in volatile liquids are considerably broadened if the concentration of the collected impurity in the solution is determined not by direct chromatographic analysis but by head-space analysis. Such a method can be accomplished both with and without the enrichment of a gas by an impurity.

The simplest case, in which a solution is directly subjected to head-space analysis, involves preliminary saturation of the solution by the investigated gas. Enrichment is not completed as the concentration of the impurities in the equilibrium gas does not differ to any significant extent from the initial concentration C_G^0. This method may be considered only as a method of sample collection that allows samples of an investigated gas to be stored over an extended time. The solvent serves as a buffer, stabilizing the concentration of a trace impurity in the gas sample by suppressing its adsorption on the container walls.

It is possible to enrich a gas with the impurity being determined if the analysis combines equilibrium concentration and head-space analysis. For this an additional operation is included, leading to a sharp drop in the initial value of the distribution coefficient of the impurity in the concentrate (increase in temperature, salting out, diluting an organic solvent with water, etc.). After processing in compliance with Equations (1.3) and (4.17), the concentration of an impurity in a gas over the obtained solution (Fig. 4.26) is related to C_G^0 by the equation

$$C_G' = C_G^0 \frac{KV_L}{K'V_L' + V_G'} \frac{f_s}{1 - FK} \qquad (4.27)$$

(primes indicate the parameters after reduction of the distribution coefficient).

This equation indicates that the degree of enriching the investigated gas by the impurities (C_G'/C_G^0) basically depends upon the relationship between the values of K and K', since the factor $f_s/(1 - FK)$ differs only slightly from unity. Also, the volume ratio V_G'/V_L' usually fluctuates within the limits 1–5. Therefore, considering the possibilities of increasing the sensitivity of the HSA method by decreasing the values of the distribution coefficients (see

210 EQUILIBRIUM CONCENTRATION OF IMPURITIES OF GASES

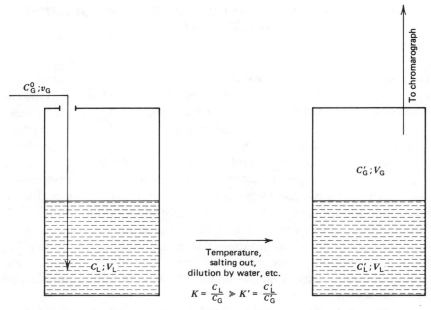

Figure 4.26. Block diagram of the determination combining equilibrium concentration and head-space analysis (see explanation in the text).

Section 1.5), the detection limit using equilibrium concentration in the volatile liquids can be lowered by one to three orders of magnitude.

Examples of measurements combining equilibrium concentration and head-space analysis include the determination of aromatic hydrocarbons in air by collecting them in water[19,20] or in acetic acid.[21]

When impurity absorption is completed using water that has been saturated with the investigated gas, the direct introduction of the solution into the chromatograph is not recommended due to low values of K (Table 4.7), which cause a decrease in the sensitivity of the analysis by more than two orders of magnitude, [Eq. (4.26)]. Therefore, after a sufficient amount of gas is purged, it is recommended that the gas in equilibrium with the obtained solution be analyzed.

Samples are collected by the saturation of 10–20 ml of purified distilled water (placed in a medical syringe), with multiple substitution of the gas phase in the syringe either by the investigated air or by purging the solution with small bubbles of air. When prolonged storage or transportation of the

4.5 COMBINATION WITH EQUILIBRIUM CONCENTRATION

selected samples to centralized laboratories is necessary, water in a glass vial is saturated with the investigated gas. In the first case, the gas in the syringe above the water is subjected to direct chromatographic determination. In the second case, the water solution is transferred from the vial into a thermostated container to equilibrate (see Section 2.2), and then the equilibrium gas phase is subjected to analysis. If the temperature of the analysis differs from the temperature of the syringe, it is necessary to correct for the change in the distribution coefficient and the volume of the gas phase. Here, the initial concentration of hydrocarbons is calculated according to the formula

$$C_G^0 = C_G' \frac{K'V_L + V_G'}{KV_L + V_G} \qquad (4.28)$$

(primes indicate parameters at the temperature of the analysis).

A distinguishing feature of this method is its simplicity and speed. The time of sample collection while purging 10–20 ml of water with the gas does not exceed 2–5 min, as v_G^{min} is only 100–130 ml. The use of such small volumes of air increases the selectivity of the analysis, since the accompanying impurities with large distribution coefficients (alcohols, carbonyl compounds, amines) do not have time to accumulate in the liquid, and their equilibrium concentration in the gas phase is extremely negligible. The method permits the determination of aromatic hydrocarbons in wet air over the concentration range of 1–50 mg/m^3 with an error of 3–6%. As such, it can be used for the analysis of exhaust gases of internal combustion engines, of the air in industrial buildings, garages, and so on.

A considerable increase in the sensitivity of the determination of aromatic hydrocarbons in air (enrichment of a gas by impurities) is achieved by concentration in acetic acid. Next, acetic acid is neutralized with a concentrated aqueous solution of caustic potassium carbonate and the gas phase in equilibrium with this solution is analyzed by the HSA method. When acetic acid is neutralized with the concentrated caustic potassium carbonate solution, the values of K sharply decrease: the values of the distribution coefficient of benzene, toluene, and m-xylene in an aqueous potassium acetate solution are much lower than in water (salting-out effect) and they are 10^3–10^4 times lower than in acetic acid. Therefore, in accordance with Equation (1.43), changing from a direct gas-chromatographic analysis of the acetic acidic concentrate to the analysis of the equilibrium vapor after neutralization of the solution allows an increase of the sensitivity of the determination by $10^3\, V_L/(K'V'_L + V'_G)$ times. The increase at $V_L/V'_L = 0.5$ and $V'_G/V'_L = 5$ (the conditions achieved in ref. 21) is almost 100-fold.

The important advantage of the method combining equilibrium concentration in acetic acid with head-space analysis is the possible determination of aromatic hydrocarbons in air with high absolute humidity (up to 23 mg/liter) at the microgram per cubic meter level. This is possible due to the increase in the analytical sensitivity allowing the use of dilute acetic acid instead of glacial acetic acid on the concentrator solvent for aromatic hydrocarbons. Due to the decrease in the K values compared with those in glacial acetic acid, the minimum volume of air necessary for equilibrium concentration of hydrocarbons in 2 ml of 80% acetic acid at 25°C is reduced from 20 liters to 6 liters. Purging this volume of air even at 100% humidity leads to the dilution of 80% acid by only 2%. Moreover, the change in K does not exceed the error of its determination (~10%). Only when samples are collected during rainy weather at temperatures over 25°C must the change in the composition of the absorbing liquid be considered. Besides that, the melting point of the 80% acid is considerably lower than that of glacial acetic acid and sample collection can be accomplished at environmental temperatures down to −7°C.

The selectivity of the equilibrium concentration method for aromatic hydrocarbons in acetic acid with neutralization of the concentrate and the analysis of the equilibrium vapor is obvious when compared with the results of adsorption concentration of atmospheric hydrocarbons using total adsorption on hydrophobic graphitized carbon black.[22] In one single sample of the concentrate on carbon black from 10 liters of air, it is possible, using gas chromatography—mass spectrometry, to detect more than 100 hydrocarbons originally present in the air. Figure 4.27 illustrates typical chromatograms of parallel samples of atmospheric impurities obtained when comparing the methods. The selectivity of the determination of the simplest aromatic hydrocarbons using equilibrium concentration is limited by two factors: (1) employing acetic acid at the stage of sample collection depletes the solution of impurities with small distribution coefficients; (2) after neutralization of the concentrate, the equilibrium gas phase is depleted of substances the distribution coefficients of which have increased more than those of the aromatic hydrocarbons (such as e.g. oxygen-containing compounds).

In addition to the possibility of stabilizing the impurity concentration of

4.5 COMBINATION WITH EQUILIBRIUM CONCENTRATION 213

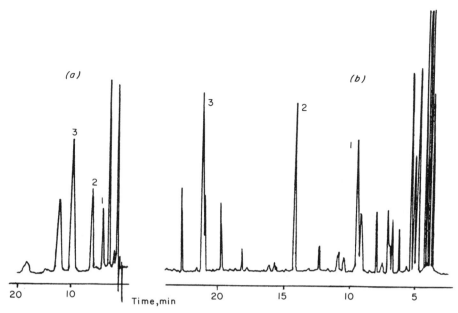

Figure 4.27. Chromatogram of hydrocarbons present in urban air. (*a*) Analysis of the equilibrium vapor over the neutralized acetic acid solution which was saturated with the air. (*b*) Analysis of the air sample using adsorption concentration on graphitized carbon black. Chromatographic conditions: ·(*a*) Column: 125 cm × 3 mm i.d., glass, containing 10% tripropionitrileamine on Chromosorb P (60–80 mesh). Column temperature: 80°C. Carrier gas (nitrogen) flow rate: 20 ml/min. Flame-ionization detector, with a full-scale response of 2×10^{-12} A. Sample size: 2 ml. Temperature of the gas sampling valve: 100°C. (*b*) Column: 50 m × 0.32 mm i.d., copper capillary, coated with dinonyl phthalate. Column temperature: programmed from 40°C at 3°C/min. Carrier gas (nitrogen) flow rate: 4 ml/min. Flame-ionization detector, with a full-scale response of 2×10^{-10} A.

the investigated gas and to the increase of the sensitivity and selectivity of the analysis, further advantages of combining equilibrium concentration with head-space analysis must be pointed out:

1. Increased accuracy of the analysis (determination of an absolute value of C_G), since sampling reproducibility of gas samples is almost an order of magnitude better than of liquid samples.
2. The possibility of automating the analysis using commercially available automatic analyzers and equipment (e.g., head-space analyzers F-42 and F-45 and sampling accessory HS-6 manufactured by Perkin-Elmer).

REFERENCES

1. L. L. Claypool and R. M. Keefer, *Proc. Amer. Soc. Hort. Sci.*, **40**, 177 (1942).
2. R. B. Sharp, *J. Agr. Eng. Res.*, **9**, 87 (1964).
3. J. Novák, V. Vašák, and J. Janák, *Anal. Chem.*, **37**, 660 (1965).
4. M. Selucky, J. Novák, and J. Janák, *J. Chromatogr.*, **28**, 285 (1967).
5. J. Gelbičová-Ružičková, J. Novák, and J. Janák, *J. Chromatogr.*, **64**, 15 (1968).
6. A. Dravnieks and B. K. Krotoszynski, *J. Gas Chromatogr.*, **6**, 144 (1968).
7. A. Dravnieks and B. K. Krotoszynski, *J. Gas Chromatogr.*, **4**, 367 (1966).
8. B. V. Ioffe, A. G. Vitenberg, and V. N. Borisov, *Zh. Analit. Khim.*, **27**, 1811 (1972).
9. A. G. Vitenberg, M. A. Kuznetsov, and B. V. Ioffe, *Dokl. Akad. Nauk SSSR*, **219**, 921 (1974).
10. A. G. Vitenberg, M. A. Kuznetsov, and B. V. Ioffe, *Zh. Analit. Khim.*, **30**, 1051 (1975).
11. V. N. Borisov, B. V. Ioffe, and A. G. Vitenberg, *Zh. Analit. Khim.*, **30**, 1289 (1975).
12. B. V. Ioffe, A. G. Vitenberg, V. N. Borisov, and M. A. Kuznetsov, *J. Chromatogr.*, **112**, 311 (1975).
13. V. V. Tsibul'skii, A. G. Vitenberg, and I. A. Khripun, *Zh. Analit. Khim.*, **33**, 1184 (1978).
14. B. V. Stolyarov and A. G. Vitenberg, Tezisy Vsesoyuznogo Nauchno-Tekhnicheskogo Soveshchaniya po Primeneniyu Gazovykh Khromatografov v Narodnom Khozyaistve, Chelyabinsk, 1977, p. 124.
15. A. G. Vitenberg, V. N. Borisov, and V. A. Isidorov, *J. Chromatogr.*, **104**, 51, (1975).
16. V. N. Borisov, A. G. Vitenberg, N. Sh. Vol'berg, V. A. Isidorov, and B. V. Stolyarov, *Tr. Gl. Geofiz. Observ. imeni A. I. Voeikova*, **1973**, (293) 83.
17. I. A. Tsibul'skaya, A. G. Vitenberg, and B. V. Ioffe, *Zh. Analit. Khim.*, **34**, 557 (1979).
18. I. A. Tsibul'skaya, A. F. Osokin, and A. G. Vitenberg, *Vest. Leningr. Univ.*, **1980**, (10) 99.
19. A. G. Vitenberg, V. V. Tsibul'skii, and I. A. Khripun, USSR patent No. 612, 171; *Byull. Izobr.* **1978** (23).
20. V. V. Tsibul'skii, I. A. Tsibul'skaya, and N. N. Yaglitskaya, *Zh. Analit. Khim.*, **34**, 1364 (1979).
21. A. G. Vitenberg and I. A. Tsibul'skaya, *Zh. Analit. Khim.*, **34**, 1830 (1979).
22. B. V. Ioffe, V. A. Isidorov, and I. G. Zenkevich, *J. Chromatogr.*, **142**, 787 (1977).

CHAPTER FIVE

Qualitative Analysis and Other HSA Applications

5.1 INDIVIDUAL AND GROUP IDENTIFICATIONS

It is not necessary to discuss methods of identification by retention parameters[1] using head-space analysis, as they are discussed in detail in the literature. They do not have any specific characteristics. However, there are three other possible approaches to qualitative analysis of the vapor phase that are important:

1. Identification of individual compounds and components of mixtures using numerical values of the gas–liquid distribution coefficients.
2. Treatment of the vapor with chemical reagents followed by gas-chromatographic monitoring of the composition change in the vapor phase.
3. Identification of liquids and solids having a complex composition by the characteristic chromatograms of the gas phase in contact with them.

The first of these utilizes the distribution of the analyzed components between the phases of a heterogeneous system outside the chromatographic column. This is the "chromato-distributive method of analysis."[2] Up to now, attention has been given primarily to heterogeneous liquid–liquid systems. Measuring distribution coefficients between gas and liquid phases by head-space analysis has been rare and its potential possibilities have not been developed. The realization of such methods of identification depends on the availability of sufficient data concerning the distribution coefficients of various compounds. The range of accurately measured gas–liquid distribution coefficients using direct methods is fairly large, covering approximately six orders of magnitude (from 10^{-2} to 10^4). At the beginning of a homologous series, the differences in physico-chemical properties of the neighboring homologs and isomers are as a rule more significant. This presents favorable prospects for the individual identification of volatile organic compounds using values of K. There is still insufficient data, and therefore the method has not yet attained wide acceptance.

The prospects of group identification by distribution coefficients remain unclear. Apparently, it will be expedient to use the correlations between the distribution coefficients in two to three different gas–liquid systems completely similar to the correlations of liquid–liquid systems.[2,3]

The second possible approach to qualitative head-space analysis (the interaction of vapors of the investigated substances with chemical reagents) was developed by Hoff and Feit.[4] It is applicable to the determination of functional groups in volatile organic compounds.

The analysis of the vapor phase allows a reduction in the sample size of the analyzed substance to several microliters and considerably simplifies the processing by the group of reagents. All operations are completed in standard 2–10 ml glass syringes. A vapor sample of the investigated substance (first diluted with air in a 0.5-liter Erlenmeyer flask with a rubber septum to a concentration of 0.1 mg/liter) is drawn into the gas syringe, and an aliquot of it is introduced into the gas chromatograph. A drop of reagent is placed on the inside wall of another syringe using a micropipet (5–10 μl for a 2-ml syringe or 25 μl for a 5-ml syringe) and spread along the surface. The syringes are connected by a short tube made of standard injection needle, as shown in Figure 5.1. In this apparatus, the vapors of the analyzed substance

Figure 5.1. Glass syringes connected by syringe needle tube for the treatment of the vapor phase with a group reagent.

pass into the syringe with the reagent. As a rule, 3 min is sufficient to complete the reaction. Once complete, the syringe containing the reaction mixture is fitted with a common injection needle, and the vapor is introduced into the chromatograph. The remaining vapor can be transferred into a syringe with another reagent and again chromatographed. A comparison between the vapor chromatograms obtained after chemical treatment and the initial chromatogram allows the establishment of the nature of the corresponding components (Fig. 5.2) by the decrease in the size of certain peaks or their complete disappearance. If the interaction with a reagent is accompanied by the formation of volatile substances, then new peaks appear on the chromatogram, which serve as an important information for identification. For instance, alcohols under the effect of an acidified solution of sodium nitrite are converted into volatile alkylnitrites. Instead of alcohol peaks, the corresponding nitrite peaks appear at much shorter retention times (Fig. 5.3).

Table 5.1 gives a list of group reagents recommended for vapor phase functional analysis by Hoff and Feit,[4] supplemented by the data from other authors' works.[5-7] In addition to the liquid reagents, solids (sodium, calcium hydride, molecular sieves) and gases (ozone, hydrogen) are also used. Sodium as a plate simply sticks to the end of the glass plunger. It was suggested that other solid reagents be placed into a glass tube 2–4 cm long with 4 mm i.d. and positioned between the syringes in place of the stainless steel needle (see Fig. 5.1).[6] Also, liquid reagents can be placed in such a tube (a microreactor) coated on the surface of an inert granular material such as Chromosorb[6].* The gaseous reagent, ozone, is obtained immediately before its use by passing an electrical discharge through a syringe filled with oxygen.[4] Comparing the results of the effect of various reagents allows, in most cases, identification of the functional groups and the characteristic structure of the sample components present in the vapor, as is evident from the classification chart[4] given in Table 5.2.

The vapor-phase group identification of volatile organic substances has been used for establishing the nature of the impurities in the ethylene recovered during production of high-pressure polyethylene[6], and for the identification of volatile substances in dairy products (kefir, yogurt,

---

*The conditions of contact with the reagent in a flow-through microreactor are better, because large volumes of the vapor phase are possible. It is necessary to complete preliminary experiments in order to check for possible changes in the gas composition due to adsorption by the support and absorption of the reagent by the solvent.

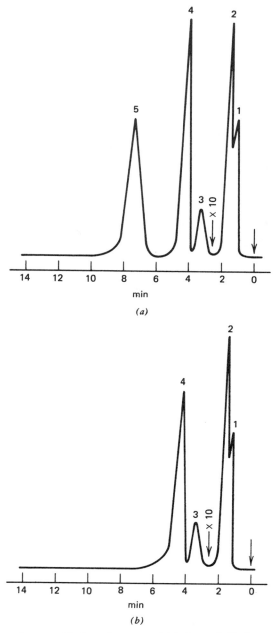

Figure 5.2. Chromatograms of hydrocarbon vapors (*a*) before and (*b*) after contact with concentrated sulfuric acid.[5] The aromatic hydrocarbons are recognized by reduction in the peak height (benzene) or total disappearance of the peak (toluene) upon sulfonation. Peaks: *1*, hexane; *2*, heptane; *3*, nonane; *4*, benzene, *5*, toluene.

Table 5.1 Reagents for Functional Vapor Phase Microanalysis

Reagent	Reagent Preparation	Result of Treatment
Metallic sodium	A thin plate is attached by pressing to the end of the syringe plunger.	Removal of all classes except ethers and hydrocarbons
Sulfuric acid	Concentrated	Removal of all classes except aromatic and paraffin hydrocarbons
80% Sulfuric acid	Mix 7 ml of concentrated sulfuric acid with 3 ml of water.	Olefins, paraffins, and aromatic hydrocarbons remain
Hydrogen	Gaseous hydrogen is drawn into the syringe, and several mg of PtO_2 is added.	Hydrogenation of unsaturated compounds
Hydrogen iodide	Heat and mix 2 ml of 90–95% phosphoric acid and several mg of potassium iodide.	Degradation of ethers[a]
Bromine water	Freshly prepared saturated aqueous solution of bromine	Bromination (removal) of unsaturated compounds
Hydroxylamine	4 g $NH_2OH \cdot HCl$ in 50 ml water	Removal of carbonyl compounds
Sodium borohydride	1 g $NaBH_4$ in 2 ml water	Conversion of carbonyl compounds into alcohols
Potassium permanganate	Saturated aqueous solution	Removal of aldehydes (the ketones remain), secondary alcohols converted into ketones
Sodium nitrite	Freshly prepared mixture of equal solution volumes of 2.5 g sodium nitrite in water and 0.1 N sulfuric acid	Alcohols converted into nitrites
Acetic anhydride	Add 2 drops of conc H_2SO_4 to 5 ml of acetic anhydride	Conversion of alcohols into esters (with subsequent addition of sodium bicarbonate)

Reagent	Conditions	Purpose
Sodium hydroxide	5% Aqueous solution	Saponification of esters
Ozone	Electrical discharge at a high voltage in a syringe filled with oxygen	Removal of unsaturated compounds by forming carbonyl compounds[b]
Hydrogen chloride	2.5 ml of conc HCl diluted with 50 ml water	Removal of amines
Water		Decrease of the concentration of water-soluble compounds in the vapor phase
Sodium arsenite	5 g $NaAsO_2$ in 50 ml water	Removal of excess ozone, regeneration of ozonides
Sodium bicarbonate	2.5 g $NaHCO_3$ in 50 ml water	Removal of acids
Calcium hydride	Grind and sieve the 0.25–0.55 mm fraction	Removal of alcohols
Mercury sulfate	20% Mercury sulfate in 20% H_2SO_4	Removal of unsaturated compounds
Silver nitrate	5% Aqueous solution	Removal of acetylene
Molecular sieve 5A	0.25–0.5 mm Fraction heated at 350° in N_2 atmosphere	Removal of paraffins and straight-chain olefins
Mercuric chloride	3% Aqueous solution	Removal of sulfurous compounds
2,4-Dinitrophenylhydrazine	0.5% Solution in H_2SO_4	Removal of carbonyl compounds
Ferric chloride	Saturated aqueous solution	Oxidation of oxycarbonyl compounds (acetone into diacetyl)
Potassium dichromate	0.5 ml of 15% aqueous solution mixed with 5 ml conc HNO_3 and 100 ml of glacial acetic acid	Oxidation of alcohols to carbonyl compounds

[a] Post-treatment with sodium bicarbonate.
[b] Sodium arsenite is added before chromatographic analysis.

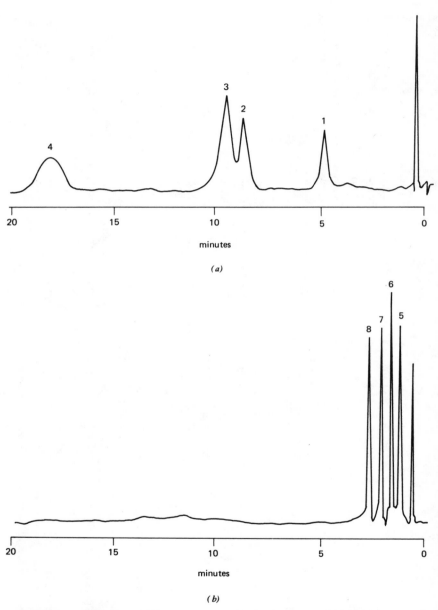

Figure 5.3. Chromatograms of the vapors of an alcohol mixture (*a*) before and (*b*) after treatment with sodium nitrite in an acidic medium.[4] Peaks: *1*, methanol; *2*, ethanol; *3*, isopropanol; *4*, propanol; *5*, methyl nitrite; *6*, ethyl nitrite; *7*, isopropyl nitrite; *8*, propyl nitrite.

Table 5.2 Classification Chart for Gas-Chromatographic Functional Analysis of Volatile Organic Compounds in the Vapor Phase[a]

Compound Class	Reagent												
	Na	H_2SO_4	H_2SO_4 (80%)	H_2	HI	Br_2	NH_2OH	$NaBH_4$	$KMnO_4$	$NaNO_2$	$(CH_3CO)_2O$	NaOH	O_3
Alcohols	D	D	D	D	D	RA	R	R	RA	DA	DA	R	RA
Aldehydes	D	D	R	D	D	RA	D	DA	D	WR	RA	R	WR
Ketones	D	D	D	WR	D	WR	D	DA	WR	WR	R	WR	WR
Esters	D	D	R	—	D	WR	—	WR	WR	WR	WR	DA	WR
Ethers	WR	D	R	WR	D	WR	—	—	WR	—	WR	—	WR
Olefins	WR	D	WR	DA	WR	DA	—	—	R	—	—	—	DA
Aromatic hydrocarbons	WR	WR	—	DA	WR	—	—	—	—	—	WR	—	—
Paraffins	WR	—	—	—	WR	—	—	—	—	—	—	—	WR

[a] D = peak disappearance; DA = peak disappearance and appearance of a new product peak; R = reduction of the peak; RA = reduction of the peak and appearance of a new peak; WR = slight reduction of the peak height; — = absence of any effect.

cheese).[7] References 5–7 also list other areas of application using head-space functional analysis in syringes. The method is applicable to compounds with a normal boiling temperature up to 200°C.

The third method mentioned at the beginning of this section concerning identification methods by the general form of the chromatograms of the volatile compounds has much wider practical applications and many more interesting possibilities. A study of the profile of a complex chromatogram is outside the scope of identification; it becomes a matter of identifying the analyzed objects. It finds wide application in medicine, biochemistry, and the chemistry of food products. This method for the characterization of liquids and solids having a complex composition needs special attention and is discussed in the following section.

5.2 STUDY OF THE CHROMATOGRAPHIC PROFILE OF COMPLEX MIXTURES—ANALYSIS OF ODORS

The term "chromatographic profile" in the analysis of complex objects means a combination of data about the number, relative position, intensity, and shape of peaks perceived as a single image. Profiles aid in the classification of the nature, origin, and peculiarities of the composition of the analyzed multicomponent system. In certain cases, one may limit the analysis to individual parts of the chromatograms, by detecting certain key components. The application of high-selectivity detectors can make a complete chromatographic separation unnecessary.

Nonchromatographic vapor-phase analysis as a means of group and individual identification is widely used in the animal world. Obviously, animal olfactory organs are fairly sensitive and selective vapor-phase analyzers. These organs are used in identification and mating searches, in sensing danger signals, and in the location of food sources.[8] Human beings, who do not have as keenly developed a sense of smell, use animals as sensitive "vapor-phase analyzers." A well-known example is the use of a tracking dog. The development of the technique of head-space gas-chromatographic analysis with the further increase in sensitivity opens a bright future for the practical application of individual and group identification of more varied objects that contain and release volatile components into the atmosphere.

5.2 CHROMATOGRAPHIC PROFILE—ANALYSIS OF ODORS

The chromatograms of many materials, drugs, parts of living organisms, and their metabolic products are so characteristic that they can serve as "fingerprints" for comparison with typical and "normal" samples even without a complete interpretation of the individual peaks. The technique of "fingerprinting" acquires even greater application in all areas of chromatography.[9,10]

Recently, various aspects concerning the use of head-space chromatograms in medicine have been intensively studied. Of note are the systematic studies of Zlatkis, Liebich and coworkers, who have developed a concentration technique for the volatile components of blood and urine on hydrophobic porous polymers. They interpret the profiles in relation to the diagnosis of certain illnesses and metabolic disorders.[11-15] An idea of the diagnostic possibilities obtained in a study of vapor-phase chromatograms can be seen in Figure 5.4. Here the volatile components in urine samples from a healthy and a diabetic person are absorbed on Tenax* and the typical chromatograms obtained are given. The most effective method for the interpretation of complex chromatograms is the combination of gas chromatography with mass spectrometry. Also, the utilization of selective detectors can be fairly useful. These detectors allow separate classes of compounds containing sulfur, nitrogen, phosphorus, or halogen to be selectively registered. Figure 5.5 shows chromatograms illustrating considerable changes in the composition of the volatile sulfur containing compounds in the urine of a diabetic person.

Literature references indicate the use of profiles of the vapor phase for the identification of drugs,[16] chemotaxonomy of grains,[17] quality control of food products,[18,19] identification of contaminants of air[20] and soil[9], contaminants, and trace identification of inflammable materials during investigation into the causes of fires.[21] In these characteristic and diverse applications of headspace analysis the accuracy and completeness of volatile component extraction is not as important as the reproducibility and high sensitivity of the chromatogram. For this purpose it is unnecessary to establish the distribution coefficients or to completely extract the volatile components. Therefore, equilibrium conditions are not required. However, strict control must be observed, with respect to experimental conditions: technical details and conditions of sample selection, processing, and chromatographic procedures. The requirement of maximum sensitivity necessitates, in most

---

*Tenax GC is a hydrophobic, thermally stable, porous organic material which is presently considered as the best sorbent for the vapor-phase analysis of aqueous solutions (see Section 3.1). Chemically it is poly(2,6-diphenyl-p-phenylene oxide).

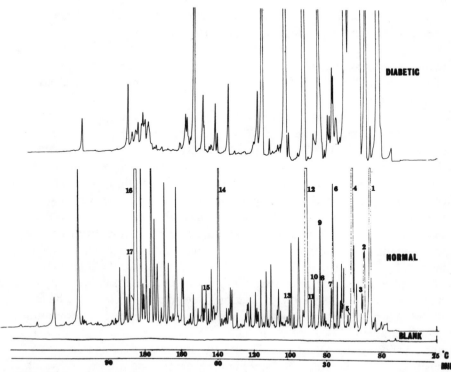

Figure 5.4. Chromatograms of the volatile organic substances present in urine samples from a diabetic and a healthy person. Peaks: *1*, butanone-2; *2*, ethanol; *3*, 3-methylbutanone-2; *4*, pentanone-2; *5*, 2,4-dimethylfuran; *6*, dimethyl disulfide; *7*, 2,3,5-trimethylfuran; *8*, 5-methylbutanone-3; *9*, pentene-2-one-3; *10*, 4-methylpentene-2-one-3 (tentative); *11*, butanol-1; *12*, heptanone-4; *13*, heptanone-2; *14*, pyrrole; *15*, benzaldehyde; *16*, carvone; *17*, piperitone.

cases, a preliminary concentration of the vapors. Moreover, this should be done using a minimum quantity of the investigated material. This is especially important in applications in medicine, physiology, and criminology. Head-space analysis using small volumes of samples usually gives unsatisfactory results and needs further improvement in the technique of concentration. One of the latest achievements is a micro technique developed in the laboratory of A. Zlatkis[22] that is used to obtain the vapor-phase "fingerprints" of 1–2 drops of water samples. A 25–200-μl sample is introduced into a 2-mm i.d. × 70 mm glass tube containing 0.3 ml of the porous hydrophilic silica gel sorbent Porasil E. The tube with the adsorbent is located in the lower part of a special apparatus called a "transevaporator"

5.2 CHROMATOGRAPHIC PROFILE—ANALYSIS OF ODORS

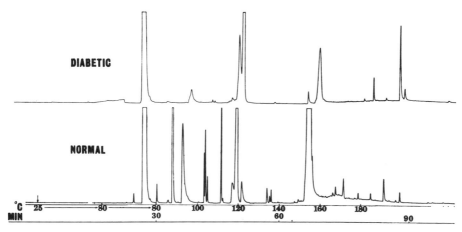

Figure 5.5. Chromatograms of the volatile sulfur compounds present in urine samples from a diabetic and a healthy person, obtained with help of a flame-photometric detector.

(Fig. 5.6) and is connected by a Teflon sleeve through a small condenser with an upper adsorbing tube filled with Tenax (1.8 ml). Helium (16 ml/min) purging for 5–10 min through this apparatus transfers the volatile substances from the tube with Porasil E into the tube with Tenax. The condenser serves to partially condense the water vapors. The tube with the Tenax is then removed and purged with helium for approximately 3 min to remove the water vapors. The volatile components are preliminarily desorbed into a primary column (30 cm × 1 mm) containing SF-96 as the stationary phase, cooled by liquid nitrogen, while the tube with Tenax is heated for 10 min up to 280°C in a slow sweep of helium (7 ml/min). The primary column is connected to the analytical capillary column, of which the first 30 cm is cooled by liquid nitrogen. The concentrate is introduced into the analytical column by rapidly heating the primary column up to 180°C during a 1-min period being purged with helium at 1.5 ml/min.

This microtechnique using vapor-phase concentration allows the collection of 20–80% of the volatile substances present in the investigated objects at concentrations as low as 5×10^{-7}%. The relative standard deviation of the peak areas is approximately 8%. This is sufficient for the characteristic profile of chromatograms of biological, medical, and food substances. In "normal" samples of this type, there are fluctuations in the component concentrations in the order of 20–50%. The shape of the vapor-phase "fingerprints" obtained by the micro technique is shown in Figure 5.7,

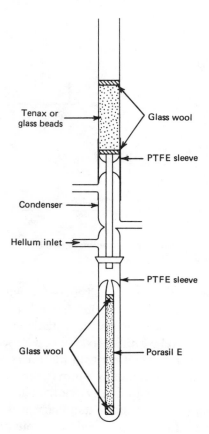

Figure 5.6. Apparatus for the concentration of the volatile components from small volumes of aqueous solutions and biomedical fluids.

which reproduces a chromatogram of the volatile components collected from one drop of blood serum of a diabetic person.

Vapor-phase concentration successfully complements the micro extraction technique and allows the less volatile components to be more completely extracted. Extraction with isopropyl chloride can be accomplished in the same apparatus (Fig. 5.6), in which a tube filled with fine glass beads (80–100 mesh) is substituted for the Tenax tube. Isopropyl chloride (0.8 ml) is placed in the lower test tube, and helium is used to push the substance through the adsorber with Porasil E and the sample into the tube with the glass beads. Soluble organics are extracted and transferred to the glass beads while most of the water and high-molecular-weight material is retained by the Porasil E. Then the adsorber with Porasil is removed, the lower part of

5.2 CHROMATOGRAPHIC PROFILE—ANALYSIS OF ODORS

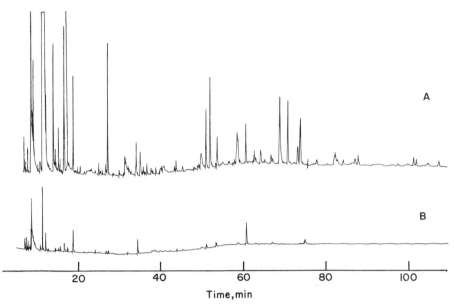

Figure 5.7. (*A*) Chromatogram of the volatile components present in the blood serum (60 μl) of a diabetic patient. (*B*) System blank.

the apparatus is heated on a water bath (50°C) and then purged for 10 min with helium to remove the excess of isopropyl chloride and traces of water. The trace organic substances transfer from the surface of the glass beads to an analytical column in the same way as from Tenax. Figure 5.8 gives chromatograms of the volatile components extracted from one drop of a brandy by concentration on Tenax and on glass beads. In the latter case, it is easily seen that the less volatile components are registered much more clearly.

The method described allow the identification of the differences between the volatile components of the blood serum of healthy people, diabetics, and those ill with influenza. The difference between brewed coffee and coffee oils of various brands can also be demonstrated. A further development of the fingerprint technique involves the more intensive use of computers for the automatic processing and comparison of complex chromatograms containing hundreds of peaks.

A special and very specific area of application of head-space analysis comprises the study of odors. Considerable attention is devoted to the

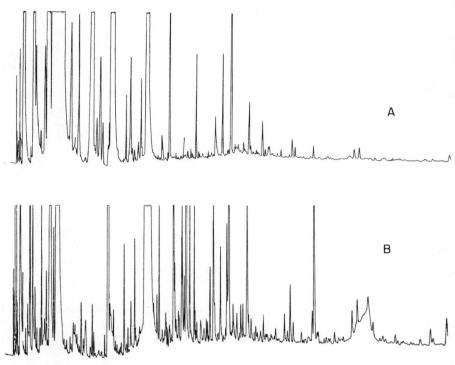

Figure 5.8. Chromatograms of the volatile components present in a brandy sample (50 µl), concentrated on (A) Tenax, and (B) glass beads.

analysis of the odors of food products, flowers, tobacco, and perfume products. There is an increasing interest in these products due to the problems of technology, storage, quality improvement, and refining.[23] The study of odors for the selection and taxonomy of fruit and essential-oil plants and for developing artificial food and imitation fragrances has even greater value.[24,25] The technique of head-space analysis is preferred for the study of odors, as smell is achieved by the olfactory organs through the gas phase. Therefore, analysis of the gas phase can give the most correct understanding of the nature and composition of the compounds forming an odor. The composition and odor of the active components obtained by other methods such as extraction, fractionation, and steam distillation differ considerably from the composition of volatile substances in the vapor phase above an investigated object. Nevertheless, these means must be used to

5.2 CHROMATOGRAPHIC PROFILE—ANALYSIS OF ODORS

extract trace high-boiling products that play an important role in the composition of odors. The characteristics of the physiology of the olfactory organs and the extremely high sensitivity and selectivity of the receptors place special requirements on the technique of head-space analysis used for such studies. The sensitivity threshold of the odors of the majority of substances is less than 1 ng/ml, and for some it is lower than 0.1 pg/ml (10^{-8}%), that is, lower than the sensitivity limit of present universal gas-chromatographic detectors. For this reason a direct analysis of the vapor phase above the object during a study of odors is relatively rarely done, and preliminary concentration of the volatile components is usually required. In individual cases it is necessary to process tens and hundreds of kilograms of the initial material. Therefore, a real danger of contaminating the negligible quantities of a concentrate by impurities from the solvents and lubricants, by oxidation products and products of other secondary reactions leading to the formation of artifacts—becomes apparent. Sample selection for odor analysis presents an independent problem, due to the above considerations, to which special attention must be given. (See, for instance, the special review by Weurman.[26]) However, the main characteristic of head-space analysis applied to odor study requires that the odorant comprise (both in number and concentration) only an insignificant fraction of the volatile components contained in the vapor phase. Therefore, problems of separation (extraction and identification of odorants against a background of substances that do not have an odor) arise. Also, there are problems of correlating the character of odors with the profile of the vapor-phase chromatograms.

Similar odors can contain compounds of different composition and structure. Thus, musk has the same odor as some cyclic ketones, some aromatic nitro compounds, and certain steroids. Negligible changes in molecular structure greatly influence the intensity and character of an odor (the odors of stereoisomers, for instance, are completely different). With such complex and ambiguous relationships between structure and odor, the determination of a component's structure and peak identification in a chromatogram necessary for recognizing known odorants are insufficient for studies of little-known objects and for determining previously unknown odoriferous substances.

While chromatographically separating the volatile components of the vapor phase while investigating the odors, one must also carry out the sensory evaluation of the odor of the eluted fractions.[27] For this reason, a

232 QUALITATIVE ANALYSIS AND OTHER HSA APPLICATIONS

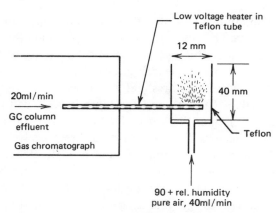

Figure 5.9. Schematic of a special outlet to a gas chromatograph for the sensory evaluation of odors.

portion of the gas current coming from the chromatographic column is diverted into an inhalation device placed parallel to the detector (Fig. 5.9). The outlet tube (Teflon or stainless steel) is heated to prevent condensation, and the carrier gas with a particular component enters a small Teflon beaker where it mixes with humid air. Humidifying decreases the initation of the mucous membrane of the nose caused by dry gases. The current distribution of the effluent and air is such that the odorant vapors do not contact the Teflon walls and hence are not retained on them. Odor evaluation is done by specially trained experts who consider a great number of factors affecting the subjective perception of the odor under strictly regulated conditions securing maximum sensitivity and accuracy. As a typical example of an investigation of this type, consider the gas-chromatographic analysis of the vapors of brewing coffee with simultaneous sensory evaluation of the odors of the volatile components.[28] The profile of the vapor-phase chromatogram of the coffee aroma (the coffee is brewed with distilled water) is given in Figure 5.10. It appears that most of the volatile components possess only shades of the coffee aroma, which are characterized by experts as "acidic," "sweet," and "burned" odors. Correctly, the "coffee" odor was noted only in the last fraction of the eluate, exiting at temperatures of approximately 230° and containing such insignificant quantities of odorants that they were not recorded by the flame-ionization detector in a concentrate collected from 50 g of coffee. This example clearly illustrates the absence of a well-defined correlation of the profiles of the volatile components with olfactometric data. However, if chromatographic separation is preceded by a release of

5.2 CHROMATOGRAPHIC PROFILE—ANALYSIS OF ODORS

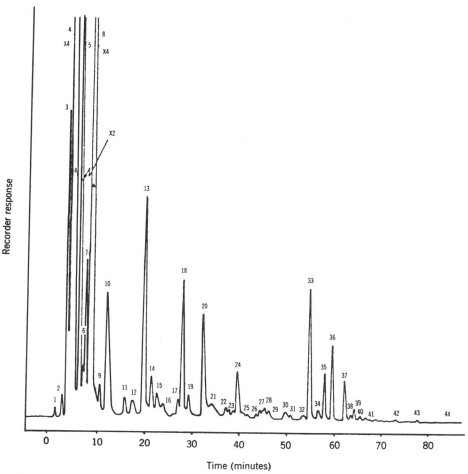

Figure 5.10. Chromatogram of the aroma of brewed coffee, after the concentration of the volatile components by sorption on Porapak. Sensory profile of the peaks: *7, 8, 10–14, 18,* and *19*, oily odor; *12, 13, 15,* and *17*, burning odor; *21–26*, mildew odor; *29–32*, odor of nuts; *36–44*, flower fragrance.

certain components of a complex odor that are chemically similar, then the chromatograms of such fractions of the odorants are more easily interpreted and very informative. Thus, characteristic profiles are obtained for the mixtures of amines, aldehydes, and sulfur compounds contained in different brands of caviar, fish, and dairy products.[24,25]

Head-space analysis has been used to study the odor of many food products, beverages, and spices.[23] Interesting special investigations also concerned the comparison of the quality of different varieties of bananas[29] and the study of the odor of apple essence.[30]

5.3 APPLICATION OF HEAD-SPACE ANALYSIS TO THE CALIBRATION AND TESTING OF GAS-CHROMATOGRAPHIC INSTRUMENTS

The principles of heterogeneous liquid–vapor equilibrium can be used to prepare two-, three- or even multicomponent mixtures with exactly known and constant composition. At first, this sounds as a trivial problem, however, it becomes sufficiently complex when one wishes to prepare very dilute gaseous mixtures containing trace amounts of the individual components. Such samples can be used as standards with known composition for the calibration and testing of the instruments used for the analysis of trace impurities.

The concentration of very dilute solutions of perfectly stable substances is not very constant and can change either during simple transfer from one container to another or upon standing. The cause of these changes lies in the sorption of the components by the container walls, rubber seal, and stoppers, or other materials. The absolute quantities of substances adsorbed per unit surface are not large, but for dilute mixtures they become comparable with the total quantity of the trace component in the container. Thus the concentration changes and related errors can reach 10–100%.[31]

Unstable concentrations of very dilute mixtures can be caused by slowly occurring chemical and photochemical reactions: oxidation, hydrolysis, and interaction with carbon dioxide and other substances. The quantities subjected to these changes in components under normal conditions are negligibly small; however, in very dilute solutions, the relative decrease in concentration is quite pronounced.

The loss of trace amounts of gaseous mixtures by sorption and chemical processes can be totally compensated for practical purposes. This can be accomplished if the gaseous mixtures are in equilibrium with a condensed phase containing the same components in much greater concentration than the gas phase and playing the role of a buffer reservoir. In this case, the total

5.3 CALIBRATION AND TESTING OF INSTRUMENTS

quantity of such components in a heterogenous system will considerably exceed their consumption by undesirable processes, and therefore, the concentration changes will be less than the allowable error. The development of heterogeneous equilibria for obtaining vapor-phase mixtures of known concentration is apparently due to Burnett and Swoboda.[32] They used the vapors of dilute aqueous solutions of ethanol and acetone for the calibration of the argon-ionization detector. The concentrations in the gas phase were calculated using literature values of Henry's constants. Later the same approach was used for the calibration of the flame-photometric detector.[33] However, the conditions and limits of the effective use of this method for the preparation of a standard sample preparation have not previously been discussed.

It is easy to show that the degree of stability of the vapor concentration can be determined by incorporating the numerical value of the distribution coefficient and the volumes of the existing phases. Due to the disappearance of amount g of a component from the gas phase of volume V_G, the concentration in the gas phase will change by the value $\Delta C_G^* = g/V_G$. However, if the equilibrium between the same gas phase and a liquid phase of volume V_L and concentration $C_L = KC_G$ shifts, then the change in the concentration of the gas phase will not be ΔC_G^*, but a lesser value ΔC_G:

$$\Delta C_G = \frac{\Delta C_G^*}{1 + K/\alpha} \qquad (5.1)$$

where $\alpha = V_G/V_L$ is the ratio of the volumes of the gas and liquid phases. Thus, the concentration decrease in the gas phase due to sorption or other reasons can be made negligibly small. For this purpose, a liquid with a sufficiently large value of K is selected, and a component solution of concentration C_L^0 is prepared such that in the equilibrium gas phase the needed concentration C_G would be retained in accordance with the formula

$$C_L^0 = (K + \alpha)C_G \qquad (5.2)$$

The concentration C_L^0 can always be made sufficiently accurate. The error in volume changes can usually be disregarded, but with large distribution coefficients the contribution to the error of C_G due to the loss of a component from the gas–liquid system will by (5.1) be negligibly small. Thus, the basic source of error regarding the concentration C_G, developing in the gas phase, will be the error in the value of K. To prepare gaseous mixtures with stable and known concentrations of trace components, exact values of the distribution coefficients must be known. The lack of reliable data concerning the distribution coefficients in dilute solutions was the apparent reason for the limited application of this simple static method of preparing standard gaseous mixtures. It is possible to use the same solution repeatedly with the total substitution of the gas phase. After the nth substitution the concentration in a gas phase will be equal to

$$C_G^n = \frac{C_L^0}{K + \alpha} \left(\frac{K}{K + \alpha}\right)^n \qquad (5.3)$$

With large values of K and similar phase volumes, the relationship $K/(K + \alpha) \approx 1$ and the differences in concentration of the successively prepared gaseous mixtures will be so negligible that they can be disregarded. The allowable value for neglecting such factors in the substitution of a gas phase above a solution can be established by the condition:

$$n \leq \left|\frac{\ln(1 - \delta)}{\ln K - \ln(K + \alpha)}\right| \qquad (5.4)$$

where δ is the permissible relative change in the concentration of the gaseous mixture ($\Delta C_G/C_G$).

It is convenient to obtain small volumes of the gaseous mixtures in equilibrium with the solutions of volatile substances in thermostated glass syringes of 50–100 ml. The use of such syringes is discussed in Chapter 2. Figure 5.11 illustrates the reproducibility of the chromatograms obtained

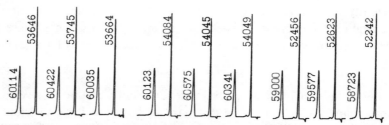

Figure 5.11. Reproducibility of chromatograms of the equilibrium vapor (head-space) above an aqueous solution (5 ml) of ethyl acetate (0.280 g/liter) and dioxane (8.29 g/liter) during a workday. Temperature: 150°C; $\alpha = 18.5$; sample size: 0.5 ml; flame ionization detector. The large peaks are due to ethyl acetate and the smaller peaks to dioxane. The numbers correspond to the peak area (integrator counts).

5.3 CALIBRATION AND TESTING OF INSTRUMENTS

using equilibrium vapor above a solution of ethyl acetate and dioxane. The relative standard deviation during a workday was ±1.1% for ethyl acetate and ±0.8% for dioxane.[37]

The possibility of using head-space analysis for metrological purposes consists in the regularity of continuous gas extraction. These regularities form the basis of the dynamic method of preparing gaseous mixtures with a strictly determined concentration. Accurate dilution of the vapor–gas mixture by purging the solution of volatile substances with the gas was recommended by Fowlis and Scott.[34] The schematic of their apparatus is shown in Figure 5.12. This apparatus was used for the calibration of chromatographic detectors and the determination of their linear range using chloroform, diisopropyl ether, toluene, chlorobenzene, and heptane, which were dissolved in squalane.[35]

Regardless of its obvious advantages, there is no information in recent literature concerning the application of this simple dynamic method for the preparation of standard gaseous mixtures. One of the possible reasons can be the ambiguity concerning the relationship of experimental conditions and the accuracy of the suggested formula. Fowlis and Scott[34] recommended

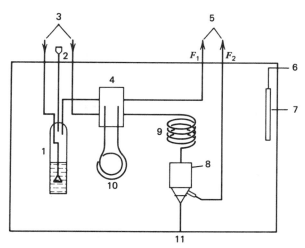

Figure 5.12. System for testing gas chromatographic detectors.[34] *1*, Container for the preparation of the gas mixture; *2*, sample introduction; *3*, argon inlet; *4*, automated gas valve; *5*, connection to flow meters; *6*, connection to the thermoregulator; *7*, resistance thermometer; *8*, argon-ionization detector; *9*, glass column; *10*, sample loop, *11*, connection to the amplifier and recorder.

the calculation of the concentration C_G of the volatile component in the gas current released from a solution by means of the formula

$$C_G = C_G^0 \exp\left[-\frac{V}{V_G + KV_L}\right] \qquad (5.5)$$

where V is the volume of the gas used in the purge, V_G is the gas volume of the container above the solution, V_L is the volume of the solution and C_G^0 is the initial gas concentration (at $V = 0$). Burnett[36] published an article almost concurrently concerning the same process to prepare a gaseous mixture through equilibrium gas extraction of the volatile substance from a solution in which a much more complex equation appeared, which in the present symbolic system has the form

$$C_G = \frac{C_G^0}{K(1 - C_G/KV_L)^2} \left[\exp(-V/KV_L) - \exp(-V/V_G)\right] \qquad (5.6)$$

The derivation of Equations (5.5) and (5.6) in the above-mentioned articles[34,36] is not clearly understood. A recent detailed study of continuous gas extraction of volatile substances from a nonvolatile solvent indicated[38,39] that Equations (5.5) and (5.6) are based on different models of the process. The formula of Fowlis and Scott (5.5) is derived assuming that thermodynamic equilibrium is instantaneously established between the gas bubbles purged through the solution, and between the solution and the gas above it of volume V_G. Thus, the gas bubbles and the gas above the solution have the same composition. On the other hand, Burnett's formula (5.6) assumes that the exchange of the volatile components does not take place between the gas volume V_G and the solution but is accomplished only in the bubbles rising through the solution.

In fact, the actual process of the exchange of the volatile components in this system is based on an intermediate between the two assumptions. Therefore, it is best to avoid the formation of considerable dead volumes of the gas above the solution. Therefore, containers with negligibly low values of V_G should be employed. In this case, the simplest relationship is valid:

$$C_G = \frac{C_L^0}{K} \left[\exp(-V/KV_L)\right] \qquad (5.7)$$

This can be used as the basis for the calculation of the dynamic method to prepare standard gas mixtures for the calibration of gas-chromatographic

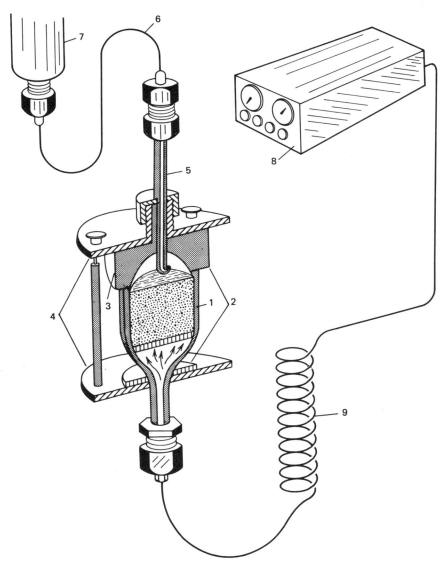

Figure 5.13. Schematic of a system for the preparation of standard gas mixtures using the dynamic method, with continuous gas extraction in a container having a small gas volume.[37] *1*, No. 16 glass filter; *2*, rubber seals; *3*, PTFE cover; *4*, braces; *5*, glass capillary; *6*, stainless steel capillary (0.2 mm i.d.); *7*, flame-ionization detector; *8*, carrier-gas (nitrogen) regulator of the gas chromatograph; *9*, copper capillary (0.2–0.5 mm i.d.).

instrumentation. The device recommended to accomplish this method is shown in Figure 5.13.

There is one more possibility to employ liquid–vapor equilibria when testing gas-chromatographic instrumentation. This consists of the use of azeotropes.[40–42] The azeotropes have identical composition in the liquid and vapor phases. This characteristic increases the accuracy of the measurements by lowering the error related to the change of the composition of the mixture by partial evaporation during the preparation and introduction of the samples into chromatograph. The problem of using the same mixture with any type of volumetric apparatus (for both gases and liquids) is simultaneously solved. Two-component mixtures of benzene and cyclohexane are convenient as standards both for testing GC instrumentation and for instructive purposes (Table 5.3). Also useful are the three-component mixtures of benzene, cyclohexane, and isobutanol, which correspond to the azeotrope composition at room temperature. The components of these mixtures are available, completely stable, nonhygroscopic, inert, and easily purified, and have similar values of the molar heat of vaporization. The similarity of the heat of vaporization results in a reduced dependence of the azeotrope composition upon temperature and makes thermostating unnecessary; this solves many practical problems. The accuracy of the determination of the relationship between the peak areas obtained using the azeotrope mixtures was 0.3% when introducing liquid and 0.1% when introducing vapor samples.

Among the various methods of using azeotropic mixtures in gas chromatography, the following are the most important:

1. Determination of the reproducibility of a gas chromatograph according to the relationship and absolute magnitude of the peak areas of the sample components.
2. Calibration of sampling devices by comparing the peak areas of a mixture introduced with the device being calibrated with the peak areas resulting from equilibrium vapor introduction by a standard sampling device.
3. Checking the operation of any type of integrator under the conditions of total or partial separation of peaks.
4. Determination of the sensitivity of detectors. In this case, the azeotrope mixture is thermostated, the vapor phase is accurately introduced, and the quantity of each of the components introduced is established by the magnitude of the partial pressure.

Table 5.3 Azeotropic Properties of the Benzene–Cyclohexane System at Temperatures near Room Temperature

Boiling Point, °C	Vapor Pressure, mm Hg	Composition, wt %		Partial Pressure, mm Hg	
		Benzene	Cyclohexane	Benzene	Cyclohexane
10	54.2	45.6	54.4	25.7	28.5
15	69	46.1	53.9	33.1	35.9
20	87	46.6	53.4	42.2	44.8
25	109	47.1	52.9	53.4	55.6

The mixture of benzene with cyclohexane, corresponding to the composition of an azeotrope at 15°C (46.1% benzene and 53.9% cyclohexane) is included in the State Register of Measures and Measuring Instruments of the USSR under the section "Standard Samples."

5.4 DETERMINATION OF IONIZATION CONSTANTS OF ORGANIC SUBSTANCES IN SOLUTIONS

Besides the analytical applications, a promising area of head-space analysis is the measurement of various physico-chemical parameters. One interesting and important possibility is the determination of equilibrium constants of chemical reactions involving volatile compounds. For this purpose, gas-chromatographic methods based upon measuring the retention parameters of the chromatographed substances before and after the introduction of nonvolatile complexing agents into the liquid phase have been used successfully. However, such dynamic methods have certain limitations related to the volatility of a solvent. Other complicating factors are the contribution of adsorption to the retention parameters and the possibility of deviation from equilibrium regarding the distribution process of the chromatographed substance in a column, which is difficult to detect. These complexities disappear if static conditions of head-space analysis are used for measuring the equilibrium distribution between the phases.[43,44]

Let us discuss the simplest case of chemical equilibrium—a reversible reaction of equimolecular quantities of components X and Y with the formation of an adduct XY

$$X + Y \rightarrow XY$$

Let one of the reagents (X) be volatile and present in the vapor of a solution. In this case, to determine the equilibrium constant,

$$K_{XY} = \frac{[X][Y]}{[XY]} \quad (5.8)$$

it is sufficient to measure the equilibrium concentrations of the volatile component C_X and C'_X in the vapors of two solutions with different concentrations of the nonvolatile reagent Y. If the total quantity of the

5.4 DETERMINATION OF IONIZATION CONSTANTS

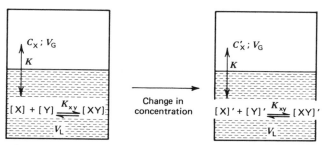

Figure 5.14. Distribution of the volatile component X between a liquid and a gas under the conditions of a reversible chemical reaction, in solution with a non-volatile reagent Y.

volatile substance in a heterogeneous liquid–gas system remains constant and the concentration of the nonvolatile reagent changes from [Y] to [Y]' (Fig. 5.14), then the equation of mass balance can be written in the form

$$C_X V_G + [X]V_L + \left(\frac{[X][Y]}{K_{XY}}\right)V_L = C'_X V_G + [X]'V_L + \left(\frac{[X]'[Y]'}{K_{XY}}\right)V_L \quad (5.9)$$

If the distribution coefficient of the volatile component remains constant in the concentration interval from [X] to [X]', that is, if

$$K = \frac{[X]}{C_X} = \frac{[X]'}{C'_X} \quad (5.10)$$

then Equation (5.9) is transformed into

$$K_{XY} = \frac{KV_L}{KV_L + V_G} \frac{C'_X[Y]' - C_X[Y]}{C_X - C'_X} \quad (5.11)$$

Depending upon the values of K_{XY} and K and also upon the method of the assignment (or determination) of the concentrations [Y] and [Y]', relationship (5.11) can be simplified and modified to be applicable to the investigation of chemical equilibria in solutions. For instance, the method extends to the complexing of volatile ligands or the electrolytic dissociation of volatile organic bases.

Stability constants of complexes with volatile ligands

$$K_{XM} = \frac{1}{K_{XY}} = \frac{[XM]}{[X][M]} \quad (5.12)$$

can be established by using a coordination substance, in amounts corresponding to an initial concentration $[M]_0$. This amount is introduced into an equilibrium heterogeneous liquid–gas system containing only a ligand, that is, $[Y] = 0$. In this case, the initial concentration C_X is lowered to C_X', and $[Y]'$ acquires the concentration value of the free coordination substance in solution, $[M]$. Then formula (5.11) with regard to the stability constant of the 1:1 coordination complex will have the form

$$K_{XM} = \frac{KV_L + V_G}{KV_L[M]} \frac{C_X - C_X'}{C_X'} \qquad (5.13)$$

The $[M]$ value differs from the initial value $[M]_0$ by the amount of the coordination substance, combined into the complex

$$[M] = [M]_0 - [XM]$$

or, in compliance with (5.10) or (5.12),

$$[M] = \frac{[M]_0}{1 + K_{XM}KC_X''} \qquad (5.14)$$

Substituting the value of $[M]$ into (5.13), we have an expression for the stability constant that considers the change in the initial concentration of the coordination substance as the result of complex formation in a solution:

$$K_{XM} = \frac{C_X - C_X'}{C_X'} \frac{KV_L + V_G}{K\{[M]_0 V_L - (C_X - C_X')(KV_L + V_G)\}} \qquad (5.15)$$

Calculation of the stability constants using (5.15) requires the determination of the absolute values of the ligand concentrations in the gas phase. Under experimental conditions of the gas-chromatographic analysis this can be accomplished with moderate accuracy. However, the determination of K_{XM} is extremely difficult. If $[M]_0 \gg [XM]$, the change in the initial concentration of the coordination substance is sufficiently small, and (5.15) is simplified to

$$K_{XM}^* = \frac{C_X - C_X'}{C_X'} \frac{KV_L + V_G}{KV_L[M]_0} \qquad (5.16)$$

The measurement of the absolute concentrations of the ligand in the gas phase can be substituted by the corresponding peak areas of the chromatogram (A_G and A_G') obtained under identical conditions. Equation (5.16) transforms as

$$K_{XM}^* = \frac{A_G - A_G'}{A_G'} \frac{KV_L + V_G}{KV_L[M]_0} \qquad (5.16a)$$

5.4 DETERMINATION OF IONIZATION CONSTANTS

Equation (5.16a) allows the determination of the stability constants on the basis of relative signal measurements from the detector. This excludes systematic determination errors in C_X and C'_X, and also excludes laborious calibration of the substance concentration in a gas from the detector signal. The criterion for the applicability of the simplified Equation (5.16a) is the condition

$$\frac{\Delta K_{XM}}{K_{XM}} \geq \frac{K_{XM} - K^*_{XM}}{K_{XM}} = \frac{(C_X - C'_X)(K + V_G/V_L)}{[M]_0} \qquad (5.17)$$

According to this, the magnitude of a systematic error due to the use of the simplified equation must not exceed the allowable determination error of the stability constant.

Thus, the determination of the stability constant of complexes with volatile ligands depends on the difficulty of measuring the relationship of the peak areas of a volatile reagent in the vapor chromatograms of two solutions. The first does not contain the coordination substance, while the second contains the coordination substance in concentration $[M]_0$.

This method can be used for the determination of stability constants of weak as well as fairly stable complexes. The analysis of expression (5.16) carried out in reference 45 indicated that at concentrations $[M]_0 \geq 0.1$ mol/liter, with volume ratio $V_G/V_L = 1$, and a 0.5% measurement error of the peak areas, it is possible to measure fairly unstable complexes having K_{XM} values of only several liters per mole. The maximum measurable values of K_{XM} are determined by introducing small quantities of the coordination substance. If the concentration $[M]_0$ developed in solution is not over 5×10^{-5} mol/liter, then it is possible to determine the values up to 10^8 liters/mol.

The lower limit of the ligand concentrations for which the investigation of complex formation is possible depends on the detection sensitivity of a ligand in the gas phase and the value of its distribution coefficient. For example, the use of the most universal flame-ionization detector relative to organic substances allows the determination of concentrations down to 10^{-3} mg/liter by direct purging of the gas phase. This method allows an error of several percent, and by Equation (1.3) one can determine the minimum concentrations of unsaturated hydrocarbons in aqueous solutions ($K \sim 0.1$), as low as 10^{-8}%. However, for ligands with high K values, for example, amines (on the order of several thousands), the minimum detectable concentrations increase to $10^{-3}-10^{-4}$%.

The applicability of the process is tested using well-investigated samples of complexes forming silver salts with olefins and aromatic hydrocarbons in aqueous solutions. Table 5.4 compares the values of the stability constants of silver complexes of four of the simplest olefins and two aromatic hydrocarbons obtained by the method discussed above with literature data. The comparison of these values is indicative of the good correlation between the results obtained by either static or dynamic methods. The larger deviation for the butylenes using the dynamic method can be explained by the low accuracy of the dynamic determination of the small distribution coefficients of olefins in volatile solvents.

The application of the above-discussed method for a study of weak reactions ($K_{XM} < 10$) is illustrated by the data of complexes of silver nitrates with benzene and toluene, which correlate well with the results of the static method. The data are based upon spectral analysis and considerably differ from the constants obtained by the dynamic method. A good correlation of the results of the static methods, based on different principles, allow one to consider these methods more reliable for unstable complexes. Apparently weak reactions in the given case do not allow the successful use of the dynamic gas-chromatographic method.

The possibilities of the determination of K_{XM} for unstable complexes are illustrated by the results of an investigation of complexes of thallium nitrate with hydrocarbons. The data given in Table 5.4 indicate that even for very weak complexes of aromatic hydrocarbons a good correlation is established,

Table 5.4 Stability Constants for Silver and Thallium Nitrate Complexes with Hydrocarbons in Aqueous Solution at 25°C

Hydrocarbon	Silver Nitrate			Thallium Nitrate (HSA Method[45])
	HSA Method[45]	Static Method	Dynamic Method[c]	
Ethylene	88.9 ± 4.0	85.3[a]	85	7.6 ± 3.1
Propylene	70.8 ± 4.2	87.2[a]	79	5.1 ± 2.3
Butylene-1	121.8 ± 8.7	119.4[a]	110	7.7 ± 3.0
cis-Butylene	65.9 ± 2.8	62.3[a]	83	4.7 ± 1.4
Benzene	2.67 ± 0.13	2.41[b]	1.58 (23.2°)	0.65 ± 0.04
Toluene	3.19 ± 0.16	2.95[b]	1.19 (23.2°)	0.93 ± 0.04

[a]K. N. Trueblood and H. J. Lucas, *J. Am. Chem. Soc.*, **74**, 1333 (1952).
[b]L. J. Andrews and R. M. Keefer, *J. Am. Chem. Soc.*, **71**, 3644 (1949).
[c]S. P. Wasik and W. Tsang, *J. Phys. Chem.*, **74**, 2970 (1970).

5.4 DETERMINATION OF IONIZATION CONSTANTS 247

as the variation of the constant does not exceed 5 relative %. However, for complexes with olefins, the error becomes 10–100% regardless of the considerably higher absolute value of K_{XM}.

Determination of the ionization constants of volatile organic bases*

The basicity of the volatile substance B, as it is known, can be characterized by the dissociation constant of the conjugate acid BH^+

$$K_{BH^+} = \frac{[B][H^+]}{[BH^+]} \quad (5.18)$$

Equation (5.18) is only a particular case of (5.8) with $X = B$, $Y = H^+$, and $XY = BH^+$. Equation (5.11) can be used to calculate K_{BH^+}. The distribution coefficient of the volatile organic bases between water and air is usually fairly large, thus $K \gg V_G/V_L$. If the value of K is unknown, the criterion for fulfilling that condition is the stability of [B] upon substitution of an equilibrium gas by a pure gas. In this case, disregarding the possibility of ionization in the gas phase, at $[H^+] > [H^+]'$ Equation (5.11) is converted into a simpler equation,

$$K_{BH^+} = \frac{(C_B/C_B')[H^+] - [H^+]'}{1 - C_B/C_B'} \quad (5.19)$$

If equal amounts of vapor at different values of solution pH are introduced into a gas chromatograph, then the relationship of concentrations of the free base in the vapors C_B/C_B' can be substituted by the ratio of the corresponding peak areas on the chromatograms A_B/A_B':

$$K_{BH^+} = \frac{(A_B/A_B')[H^+] - [H^+]'}{1 - A_B/A_B'} \quad (5.20)$$

Thus, the determination of the basicity constants of volatile organic compounds is reduced to establishing the relationship of the peak areas of free bases in the chromatograms of the vapors of solutions having different values of pH. The pH value can be measured directly using a hydrogen electrode or simply established by using buffer solutions of exactly known pH. The method employs the introduction of a certain (equal) quantity of the investigated substance containing the bases into a certain volume of

---

*See references 43, 44, 46, and 47.

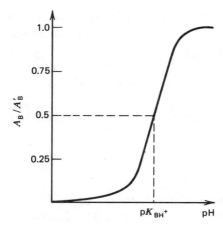

Figure 5.15. Dependence of the ratio A_B/A_B' of a base in the vapor being investigated upon the acidity of the solution.

buffer solutions of different pH. Equal-volume vapor samples of these solutions are chromatographed, and K_{BH^+} is calculated using Equation (5.20). The ionization constant can also be determined graphically by plotting A_B/A_B' of the investigated base in the vapors against the pH of the solution. In this case pK_{BH^+} is equal to pH at $A_B/A_B' = 0.5$ (Fig. 5.15).

It should be especially noted that the high sensitivity and selectivity of the gas-chromatographic analysis allow the determination of the ionization constants of bases present in complex mixtures. These systems can be composed of substances of a very different nature and different volatility, and the total qualitative and quantitative composition of these mixtures can remain unknown. However, to obtain sufficiently accurate results, it is very important to select the pH of the investigated solutions so that the peak areas A_B and A_B' can be determined with maximum accuracy. The working interval of pH lies in the area of maximum sensitivity of the area A_B' to the change of pH'.

The analysis of the dependence of the error on experimental conditions[47] in the determination of K_{BH^+} indicates that the minimum error in measuring A_B is achieved if the acidity of a solution is selected such that pH' > (pK_{BH^+} + 2). Here the areas or peak heights will be maximal and practically will be independent on the pH. The desirable pH values are between (pK_{BH^+} − 1) and (pK_{BH^+} + 0.25). The pK_{BH^+} of a great majority of volatile organic bases (amines, hydrazines, nitrogen-containing heterocyclics) lie in the range of 4–11. This entire range can be encompassed by four or five buffer solutions of pH from 3 to 12. These four or five solutions must

5.4 DETERMINATION OF IONIZATION CONSTANTS

be used for the determination of the basicity constants by the discussed method.

The results of such determinations, given in Table 5.5, indicate that the accuracy achieved in the determination of pK_{BH^+} is 0.02–0.15 units of pK. This is fully sufficient for the purposes of qualitative analysis. The method is also applicable to those cases in which the known methods for the determination of the basicity constants cannot be used.

Of greater practical interest for analytical purposes is the determination of pK_{BH^+} of bases in complex mixtures of unknown qualitative and quantitative composition. These values can be used for group and individual identification, as well as for the classification of organic substances by basicity. Formula (5.20) does not contain the concentration values of the investigated bases. Due to this, the presence of other substances and their properties in a mixture becomes unimportant.

The possibility of measuring pK_{BH^+} of the bases present in mixtures of a complex composition is discussed in references 46 and 47, using the mixture of four aliphatic amines, pyridine, and piperidine as an example. Ethanol was added as the standard substance, which remained neutral in the pH interval used. Figure 5.16 illustrates chromatograms of the vapors above the solutions of an investigated mixture of varying pH. These illustrate the quality of separation of the mixture and the reduction of the peak areas of the corresponding bases upon increasing the acidity of the solution.

The data obtained in this case (Table 5.6) indicate that the absolute error does not exceed 0.25 pK units and differ little from the same value for the individual bases. The error in the final result obtained using chromatographic peak heights is usually somewhat larger than when using peak areas. This fact can be caused by the asymmetry of peaks of the nitrous bases. Such accuracy is fully sufficient for group identification purposes.

Head-space analysis allows the determination of the ionization constants even in the event that the analyzed substance is not separated in the chromatogram from other components of the mixture. If the base and the substance not exhibiting acid-base properties appear as one peak, equal amounts of vapor are subjected to analysis and the distribution coefficient of the free base is much greater than the ratio V_G/V_L, then the ionization constant can be calculated by using the formula[8,9]

$$K_{BH^+} = \frac{[H^+](A_m/A'_m - A_n/A'_m) - [H^+]'(1 - A_n/A'_m)}{1 - A_m/A'_m} \quad (5.21)$$

Table 5.5 Results for the Determination of the Values of pK_{BH^+} for Individual Bases Using Head-Space Analysis[3]

| | Solution Acidity | | pK_{BH^+} Values | | | Absolute Error in Determining pK_{BH^+} | |
| | | | Using Equation (5.20) | | | | |
Base	pH	pH′	Computed from Peak Areas (1)	Computed from Peak Heights (2)	Literature Data (3)	Using Peak Areas (1) − (3)	Using Peak Heights (2) − (3)
Trimethylamine	8.82	10.10	9.73	9.93	9.80[a]	−0.07	+0.13
Triethylamine	10.53	13.0	10.83	10.97	10.72[b]	+0.11	+0.25
Diisopropylamine	11.27	11.58	11.18	11.23	11.20[c]	−0.02	+0.03
Pyridine	4.45	6.80	5.40	5.43	5.25[c]	+0.15	+0.18
Aniline	4.45	6.15	4.71	4.74	4.61[c]	+0.10	+0.13
N-Diethylaniline	5.42	6.80	6.66	6.60	6.56[a]	+0.10	+0.04

[a] A. Weissberger (Ed.), *Elucidation of Structures by Physical and Chemical Methods*, Part I, Interscience, New York, 1963.
[b] F. M. Jones and E. M. Arnett, in *Progress in Physical Organic Chemistry*, Vol. 11 (A. Streitwieser, and R. W. Taft, Eds.), Wiley, New York, 1974, pp. 292–295.
[c] J. A. Riddick and W. B. Bunder, *Organic Solvents, Physical Properties and Methods of Purification*, 3rd ed., Wiley International, New York, 1970.

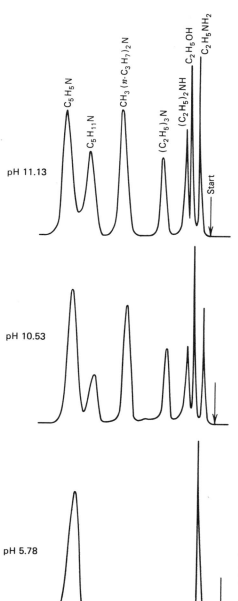

Figure 5.16. Chromatograms of the headspace above the solutions of nitrogen bases and ethanol (internal standard) at various pH values of the solution. The total concentration of mono- di- and triethylamines, methyl-dipropylamine, pyridine and piperidine is 0.3% and that of ethanol is 0.05%. Chromatographic conditions: Column: 200 cm × 3 mm i.d. glass, containing 10% polyoxyethylene 100 on Chromosorb W (0.20–0.25 mm fraction); Column temperature: 100°C; Carrier gas (argon) flow rate: 80 ml/min; flame-ionization detector.

Table 5.6 Results for the Gas-Chromatographic Determination of pK_{BH^+} Values for Bases in Mixtures[46,a]

Compound	Solution Acidity		Data from Literature (1)	pK_{BH^+}		Absolute Error	
	pH	pH'		Computed from Peak Areas (2)	Computed from Peak Heights (3)	Using Peak Areas (2) – (1)	Using Peak Heights (3) – (1)
Triethylamine	10.53	13.0	10.72[b]	10.97 10.90[d]	10.98 10.94[d]	+0.25 +0.18	+0.26 +0.22
Pyridine	5.78	8.80	5.25[c]	5.47 5.45[d]	5.43 5.41[d]	+0.22 +0.20	+0.18 +0.16
Methyl-di-n-propylamine	10.53	13.0	—	10.64	10.63	—	—
Ethylamine	10.53	13.0	10.68[b]	10.88	10.92	+0.20	+0.24
Diethylamine	10.53	13.0	11.02[b]	11.23	11.30	+0.21	+0.28
Piperidine	10.53	13.0	11.12[b]	11.24	11.28	+0.12	+0.16

[a]The gas-chromatographic conditions of the analysis of the mixture are given in Figure 5.16.
[b]F. M. Jones and E. M. Arnet, in *Progress in Physical Organic Chemistry*, Vol. 11 (A. Streitwieser and R. W. Taft, Eds.), Wiley, New York, 1974, pp. 292–295.
[c]J. A. Reddick and W. B. Bunder, *Organic Solvents, Physical Properties and Methods of Purification*, 3rd ed., Wiley International, New York, 1970.
[d]This was established from the analysis of triethylamine-acetone and pyridine-toluene systems, which produced an inseparable peak in the chromatogram. The values of A_n for acetone and toluene were established at pH 6.2 and 1.7, respectively.

where A_m and A'_m are the peak areas or heights in the gas phases in equilibrium with solutions having hydrogen ion concentrations of $[H^+]$ and $[H^+]'$, A_n is the peak area or height of a substance that does not exhibit acid-base properties (being determined by analyzing the gas phase above the solution upon strong acidification).

Formula (5.21) was verified using samples of pyridine-toluene and triethylamine-acetone mixtures (Table 5.6). The values of pK_{BH^+} obtained for pyridine and triethylamine and for the individual peak of a base and the base not separated from a neutral substance differ little. When determining pK_{BH^+} of a base appearing as an inseparable peak with the substance not exhibiting acid-base properties, the error depends not only on the values included in Equation (5.20), but also on the error in the measurement of the peak area of the neutral substance. It also depends on the fraction of the substance in the mixture of the two components in the gas phase. The larger the fraction, the less accurate the measurement.

In this case, the value of pK_{BH^+} can be determined graphically by plotting A_m/A'_m against the pH of the investigated solutions. Here, pK_{BH^+} will be equal to pH at $A_m/A'_m = (1 + A_n/A'_m)/2$. This gives a value of 10.85 for ionization constant measured for triethylamine (Fig. 5.17), which differs from the value obtained from Equation (5.21) by 0.05 pK unit.

Besides the applications already mentioned, the suggested method can be used for the analysis of compounds giving nonseparated peaks if only one of

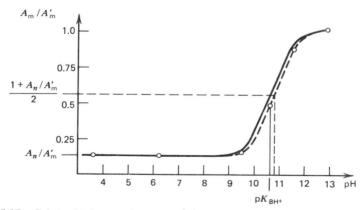

Figure 5.17. Relationship between the ratio A_m/A'_m of the vapors of triethylamine and acetone and the pH of the solution. (for the inseparable peak, $A_m/A'_m = 0.138$). The solid line is the theoretical curve [eq. (5.15)]; the broken line represents the experimentally measured data.

254 QUALITATIVE ANALYSIS AND OTHER HSA APPLICATIONS

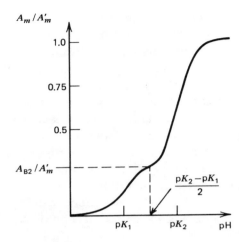

Figure 5.18. Relationship between the ratio A_m/A'_m of the vapor mixture of nonseparated bases and the pH of the solution for the case when the values of pK_{BH^+} differ by 2.5.

the two substances exhibits acid–base properties. For the determination of the inseparable peaks of a base and a neutral substance it is necessary to analyze the vapors above a sufficiently basic and acidic solution of the investigated sample (Fig. 5.17). When the accuracy of peak area or height measurement is 1%, a sharp inflection, indicating the presence of two inseparable bases, is observed at $(pK_2 - pK_1) > 2.5$ (Fig. 5.18). This considerable difference in the pK values of two components giving one peak, necessary for establishing an inflection on a curve, limits the possibilities of this method to establish the individual peaks of bases belonging to different classes. For example the HSA method can be used in the case of overlapping the pyridine peak by aliphatic amines.

5.5 DETERMINATION OF ACTIVITY COEFFICIENTS

As noted in Chapter 1, knowledge of the numerical values of the gas–liquid distribution coefficients allows the determination of the activity coefficients of volatile substances in solution. Due to this, the high sensitivity of gas-chromatographic head-space analysis becomes very important. Very dilute solutions can be used and the limiting values γ^∞ can be determined. These values are of considerable theoretical and practical interest and are difficult

5.5 DETERMINATION OF ACTIVITY COEFFICIENTS

to obtain with other methods. However, the method of continuous extraction of the volatile components in solution by an inert gas flow has broad applicability here. This was used for the first time in the determination of activity coefficients in 1977.[48]

In the case of nonvolatile solvents, one may use Equation (1.37), which upon substituting the ratio A_L^0/A_L by the ratio A_G^0/A_G and combining the resulting equation with (1.15), gives

$$\gamma^\infty = \frac{RTd_L V_L}{p^0 M v_G} \ln \frac{A_0}{A} \qquad (5.22)$$

or

$$\gamma^\infty = \frac{RT}{p^0 v_G} N \ln \frac{A_0}{A} \qquad (5.23)$$

where N represents the number of moles of the solvent.

The derivation of the initial relationships assumed that the solutions are so dilute that the constants d, and M are practically the same as those of the pure solvent, and the distribution coefficient does not depend on the concentration. However, in the case of the determination of very high activity coefficients (reaching several hundred units) one must consider that formulas (5.22) and (5.23) are only approximate. Here, even in dilute solutions, corrections must be introduced, the details of which are discussed in reference 49.

Investigation of the method of determining the activity coefficients by means of gas extraction and head-space analysis[48,50] indicated that the reproducibility of the data is considerably better than of those methods which determine the activity coefficients from retention parameters. Experiments were carried out with volatile solutions of paraffinic, naphthenic, and aromatic hydrocarbons in phenol, furfural, and dimethylformamide,[50] as well as of hexane and benzene in methylpyrrolidone, dimethyl sulfoxide, hexadecane, decalin, aniline, nitrobenzene, acetophenone, and ethylene glycol.[48]

Standard deviation of the results between the two compared methods was 6%[48] or 10–20%,[50] with the exception of one case of very high activity coefficient (hexane in diethylene glycol).

The method for the determination of activity coefficients using inert gas extraction of the solutions requires neither preliminary calibration of the chromatographic detectors nor laborious preparation of the instrumentation. It is characterized in the literature as the most accurate and attractive method, due to its simplicity, reliability, and possible application to volatile solvents.

5.6 DETERMINATION OF THE MOLECULAR WEIGHT

The classical methods for the determination of the molecular weight (cryoscopy and ebulliometry) are ultimately based on Raoult's law, which establishes the relationship between solvent vapor pressure depression and the mole fraction of the dissolved solute. The determination of molecular weight by direct vapor pressure measurements of solutions was not practical, due to the comparative complexity and difficulty of the measurement. However, the gas-chromatographic analysis of the equilibrium vapor phase allows a simple and sufficiently accurate way to determine the ratio of the vapor pressure p of a solution to the vapor pressure p_0 of the pure solvent. Within the limits of the linear range of the detector, this ratio is equal to the ratio of the chromatographic peak areas (or heights) using equal-volume samples of the vapors of the solution and solvent:

$$\frac{p}{p_0} = \frac{A}{A_0} \quad \text{and} \quad \frac{p_0 - p}{p_0} = \frac{A_0 - A}{A_0} = x$$

It is easy to derive the formula

$$M = \frac{A}{A_0 - A} \frac{m}{w} M_s \qquad (5.24)$$

for the determination of the molecular weight M of the dissolved solute given the data of the head-space analysis of a solution consisting of m grams of the substance in w grams of a solvent of molecular weight M_s.

Determinations under the simplest experimental conditions (introducing the head-space of a solution and of the solvent at room temperature from a flask into the injection port of a gas chromatograph using a gas-tight syringe) showed that substances having a molecular weight between 80 and 350 exhibited errors not exceeding 3% and averaging 1.5%.[51]

Especially convenient was the use of differential chromatography developed by Zhukhovitskii and Turkeltaub. In this method the solvent vapors

5.7 APPLICATION IN MICROBIOLOGY AND MEDICAL DIAGNOSTICS

are continuously passed into the chromatograph, functioning as the carrier gas, and the solution vapors are introduced periodically.[52] The height differences corresponding to the vapor pressure depression are directly measured as steps on the chromatogram. For solutions of C_7–C_{16} hydrocarbons and petroleum fractions in hexane or carbon tetrachloride, the error for single measurements did not exceed 2%.

The use of interval chromatography allows the determination of the average molecular weight of liquid mixtures containing components which are more volatile than the solvent. However, the construction of the apparatus is more complex and in addition to two four-way valves includes three- and six-way valves.[52]

The basic error in the determination of molecular weight by head-space analysis occurs in the measurement of the difference $A_0 - A$, which must be sufficiently large and therefore, it is necessary to work with comparatively concentrated solutions. In the investigations cited,[51,52] 10–20% solutions were used in quantities of 10 g. Therefore, the investigated substance was used in fairly large amounts (approximately 1 g and more). This is admittedly a considerable shortcoming. Also, one must remember that the method assumes compliance with Raoult's law up to the indicated high concentrations. Systems become practically inapplicable when sufficiently strong deviations from ideality occur. Applying the method to more diverse substances will require more accurate chromatographic measurements.

5.7 HEAD-SPACE ANALYSIS IN MICROBIOLOGY AND MEDICAL DIAGNOSTICS

One of the latest and a very promising area of the application of head-space analysis is the investigation of the volatile metabolites of microorganisms (fatty acids, alcohols, amines, and simplest carbonyl and sulfur compounds). As early as the beginning of the 1970s it was found that information about the composition of volatile metabolites is very useful for the chemical taxonomy of bacteria, viruses, and fungi and for the diagnosis of diseases caused by them. Gas-chromatographic analysis of liquid extracts, culture media, and clinical material has been fairly widely used in microbiological laboratories.[53-56] However, head-space analysis may be considered to be a more advisable method for the determination of traces of volatile components in specimens of this type. This procedure is safer and free of the use of inflammable solvents. This is an important consideration in microbiological and clinical laboratories. It is not necessary to inject nonvolatile and readily decomposed substances into the chromatograph. Moreover, it is possible to detect compounds masked with broad solvent peaks in the chromatograms of extracts. Chromatographic vials for head-space analysis in which volatile substances are distributed between the sample investigated and the head-space can be used to transport chemical material, and in vitro experiments they can be used to grow bacteria and fungi. The possibility of determining the metabolitic products of living cultures is also an important advantage. This allows the study of the composition of volatile substances during the growth of microbial flora under anaerobic conditions. It is also of great importance for series analyses that the existing automated head-space analyzers and special devices described in Chapter 2 can be used.

Microbiological applications of HSA have a great future. This is shown by the fact that a considerable number of the papers presented at the International Symposium on the Use of Chromatography in Microbiology (Lund, Sweden, October 1976) were devoted to this subject.

In reviews[54-57] published in recent years and in original papers[58-62], the greatest attention is devoted to the use of head-space analysis for the identification of anaerobic bacteria and the diagnosis of infectious disease. Most papers deal with the analysis of volatile fatty acids,[57,59,62] alcohols,[58,59,61] amines,[58] and dimethyl disulfide.[60,61]

Unlike many bacterial infections, diseases involving anaerobic bacteria are usually of endogenic origin, that is the microorganisms involved are part of the normal flora of the body. In normal circumstances they are not pathogenic, but if the balance of health is disturbed, then auto-infection may occur. Factors predisposing to such infection include surgery or other trauma, malignancy, diabetes, and therapy with antibiotics and x-rays, etc. The diagnosis of anaerobic infections and the testing of the efficiency of drugs are based mainly on clinical evidence. The results of bacteriological analysis are unreliable,[57] and the analysis may require several days. Hence, the physician is often forced to make a diagnosis and prescribe therapy with antibiotics on the basis of the appearance and site of the infection and the history of the patient without much support from the laboratory.

The application of head-space analysis to the diagnosis of infection has changed this situation. In the mid-1970s, several researchers demonstrated

5.7 APPLICATION IN MICROBIOLOGY AND MEDICAL DIAGNOSTICS

that the presence of volatile fatty acids with a carbon chain length of C_3 or longer in pus or wound drainage fluids represents a reliable evidence of anaerobic infection. Pus collected from infections of aerobic origin contains either no volatile acids or only acetic acid.

All characteristic volatile fatty acids (C_2–C_7 including isomers) are readily separated as symmetrical peaks on a column containing polar stationary phases such as: polyethylene glycol 20M or SP 1000. Chromosorb W pretreated with orthophosphoric acid is most often used as the solid support. The analysis time including the time for the establishment of an equilibrium does not exceed 1 hr. In order to improve the limit of detection of fatty acids, the specimen of biological material is heated to 90–95°C, acidified with sulfuric acid, and saturated with sodium chloride, ammonium, or magnesium sulfate.[57,62] Taking into account the background content of volatile substances in normal fluids of a healthy organism, it is possible to determine fatty acids at concentrations of 10–100 mg/liter and reliably reveal the presence of anaerobic infection in the specimen. This sensitivity coupled with good reproducibility makes it possible to monitor the changes in fatty acid levels in clinical material that occur during antibacterial therapy. The reliability of gas chromatographic head-space detection of anaerobic infection is fairly high. According to Taylor's data,[57] comparative investigation of 71 specimens of pus and wound exudates from sites apt to yield anaerobic bacteria (soft tissue abscesses, the gastrointestinal tract, and the female genital tract) showed agreement between HSA and anaerobic culture analysis in more than 90% of the cases. Moreover, the anaerobic infection was established by head-space analysis even when the patients had received concurrent antimicrobial therapy that would have greatly reduced the chances of successful isolation of the causative bacteria by traditional methods. The high reliability of the diagnostics of anaerobic empyema has been demonstrated.[56] Of 32 specimens of pleural fluid in which bacteriological analysis confirmed anaerobic infection, in only one case were volatile fatty acids not detected.

An even more detailed classification of anaerobes is possible: classification according to the spectra of fatty acids including higher homologs and also oxy- and dibasic acids in the form of volatile esters. However, direct analysis of chemical material does not always allow the diagnosis of the disease, owing to high background content of volatile substances, the difference in the pathogenicity or virulence of microorganisms, the existence of mixed infections, and the effect of therapeutic agents. In these cases microbes can be identified after additional cultivation in vitro. Analysis of an enriched medium under standard conditions gives a constant and well-reproducible composition of fatty acids. Owing to the high sensitivity of gas chromatography, it is possible to minimize the incubation period and to obtain results in a few hours.

Microorganisms can also be identified from the presence of specific volatile substances in the specimen, formed as a result of the metabolism or degeneration of bacterial cells. These are the most precise and reliable indications requiring no quantitative analysis and permit the use of relatively simple instrumentation. In some cases the distinguishing parameters may be different ratios of the same characteristic substances or different chromatographic profiles (see Section 5.2). These criteria require quantitative analysis. The higher its precision, the smaller are the differences in component ratio on the chromatogram that can be used for diagnostic purposes. Hence, standardization and perfection of the methods of specimen preparation for analysis and the analytical procedure are very important for the development of microbiological applications of head-space analysis.

Another field of application is the diagnosis of urinary tract infections. Urine is the most common type of specimen dealt with in the general hospital situation and there is usually a high percentage of uninfected samples. Head-space analysis makes it possible to reject those samples and to leave for further study only those liable to yield significant numbers of bacteria. *Escherichia coli* and *Proteus mirabilis* account for 95% of all urinary tract infections. Both these microorganisms can be detected in urine by head-space analysis, the former from the amount of ethanol formed by fermentation of the carbohydrate arabinose and the latter from the presence of dimethyl disulfide from the metabolism of methionine. Gas-chromatographic analysis of urine also permits detection of infections characterized by an overgrowth of the bacterial flora of the small intestine that leads to severe disease of the digestive organs. The existing tests for the degree of bacterial colonization in the small intestine are relatively complex and require the intubation of the patient. Head-space analysis makes it possible to obtain indirect data on the bacterial activity of digestive organs from the presence of phenol and *p*-cresol in urine, end products of tyrosine metabolism. Anaerobes yield *p*-cresol, whereas aerobes and facultative anaerobes produce phenol. Normal daily output in urine ranges from 7 to 12 mg/day phenol and from 45 to 60 mg/day *p*-cresol. Their ratio of approximately 6:1 (*p*-cresol:phenol) reflects the greater role of anaerobic bacterial metabolism in the healthy digestive tract. Under the conditions of overgrowth of bacterial flora in the small intestine, the total volatile phenols excreted are greatly increased, and the p-cresol/phenol ratio drops to 2 : 1. The rise in the amount of phenol excreted in the urine may be indicative of the increase in the number of enterobacteria colonizing the small intestine.

Quantitative head-space analysis of phenols can be carried out on a packed column similar to that used for the separation of volatile fatty acids.

The possibilities of using this method are by no means limited to these examples. Fungi can yield a broad spectrum of volatile substances. For example, by studying the process of their growth on various media, Norrman[63] has shown that head-space analysis of the metabolism products of Dupodascus aggregates on leucine makes it possible to record about 20 esters and simplest alcohols. Also promising is the use of head-space analysis in the medical industry for testing the sterility of instruments, water, and air, and for monitoring the bacterial infection of preparations and drugs. Hence, in the near future one can expect further developments in the microbiological applications of head-space analysis.

REFERENCES

1. M. S. Vigdergauz, L. V. Semenchenko, V. A. Ezrets, and Yu. N. Bogoslovskii, *Kachestvennyi Gazokhromatograficheskii Analiz*, "Nauka," Moscow, 1978, 243 pp.
2. V. G. Berezkin, V. D. Loshchilova, A. G. Pankov, and V. D. Yagodovskii, *Khromatoraspredelitel'nyi Metod*, "Nauka," Moscow, 1976, 112 pp.
3. Z. St. Dimitrova, "Ispol'zovanie Fazovykh Ravnovesii v Gazokhromatograficheskom Analize Aminov." Kand. Diss. Leningrad State University, 1976, 133 pp.
4. J. E. Hoff and E. D. Feit, *Anal. Chem.*, **36**, 1002, (1964).
5. E. Kaminski, E. Wasowicz, and R. Przibylski, *Chem. Analy*, **17**, 1307 (1972).
6. E. E. Kugucheva, A. V. Alekseeva, in *Gazovaya Khromatografiya*, NIITEKhim, Moscow, Vol. 13, (1970), p. 64.
7. V. Palo, *Chromatographia*, **4**, 55 (1971).
8. M. Barbier, *Vvedenie v Khimicheskuyu Ekologiyu*, "Mir," Moscow, 1978, 229 pp; M. Barbier, *Introduction à l'Ecologie Chimique*, P. Masson, Paris, 1976.
9. R. Jeltes, *J. Chromatogr. Sci.*, **12**, 599 (1974).

10. I. R. Politzer, B. J. Dowty, and J. L. Laseter, *Clin. Chem.*, **22**, 1775 (1976).
11. A. Zlatkis, W. Bertsch, H. A. Lichtenstein, A. Tishbee, F. Shunbo, H. M. Liebich, A. M. Coscia, and N. Fleischer, *Anal. Chem.*, **45**, 763 (1973).
12. H. M. Liebich and O. Al-Babbili, *J. Chromatogr.*, **112**, 539 (1975).
13. H. M. Liebich and J. Wöll, *J. Chromatogr.*, **142**, 505 (1977).
14. H. M. Liebich, O. Al-Babbili, A. Zlatkis, and K. Kim, *Clin. Chem.*, **21**, 1294 (1975).
15. A. Zlatkis, H. A. Lichtenstein, and A. Tishbee, *Chromatographia*, **6**, 67 (1973).
16. L. V. S. Hood and G. T. Barry, *J. Chromatogr.*, **166**, 499 (1978).
17. F. W. Hougen, M. A. Quilliam, and W. A. Curran, *J. Agr. Food Chem.*, **19**, 182 (1971).
18. M. Koijuma, *J. Food Sci. Technol.*, **20**, 316 (1973); *R. Zh. Khim.*, **1974**, 3R307.
19. D. Nurok, J. W. Anderson, and Z. Zlatkis, *Chromatographia*, **11**, 188 (1978).
20. W. Bertsch, R. C. Chang, and A. Zlatkis, *J. Chromatogr. Sci.*, **12**, 175 (1974).
21. G. Paulig, in *Vorträge zum 2. Intern. Colloquium über die gas-chromatographische Dampfraumanalyse*, Überlingen, 1978.
22. K. Y. Lee, D. Nurok, and Z. Zlatkis, *J. Chromatogr.*, **158**, 377 (1978).
23. G. Charalambous (Ed.), *Analysis of Foods and Beverages. Headspace Techniques*, Academic Press, New York, 1978, 394 pp.
24. R. V. Golovnya, *Usp. Khim.*, **45**, 1895 (1976).
25. R. V. Golovnya, "Issledovanie Komponentov Zapakha Pishchevykh Produktov." Doct. Sci. Diss. Inst. Organo-Element Compounds, USSR Acad. Sci., Moscow, 1973.
26. C. Weurman, "Sampling in Airborne Odorant," in *Human Responses to Environmental Odors*, Academic Press, New York, 1974, p. 263.
27. A. Dravnieks and A. O'Donnell, *J. Agr. Food Chem.*, **19**, 1049 (1971).
28. C. G. Tassan and G. F. Russel, *J. Food Sci.*, **39**, 64 (1974).
29. I. McCarthy, J. K. Palmer, C. P. Shaw, and E. E. Anderson, *J. Food Sci.*, **28**, 379 (1963).
30. R. G. Buttery and H. K. Burr, *Food Technol.*, **20**, 166 (1966).
31. H. G. Maier, *J. Chromatogr.*, **50**, 329 (1970).
32. M. G. Burnett and P. A. T. Swoboda, *Anal. Chem.*, **34**, 1162 (1962).
33. T. G. Field and J. B. Gibbert, *Anal. Chem.*, **38**, 628 (1966).
34. I. A. Fowlis and R. P. W. Scott, *J. Chromatogr.*, **11**, 1 (1963).
35. I. A. Fowlis, R. G. Maggs, and R. P. W. Scott, *J. Chromatogr.*, **15**, 471 (1964).
36. M. G. Burnett, *Anal. Chem.*, **35**, 1567 (1963).
37. A. G. Vitenberg and M. I. Kostkina, *Zh. Analit. Khim.*, **34**, 1800 (1979).
38. A. G. Vitenberg and M. I. Kostkina, *Vestn. Leningr. Univ.* **1980**, (4), 110.
39. A. G. Vitenberg and M. I. Kostkina, *Zh. Analit. Khim.*, **35**, 539 (1980).

40. B. V. Ioffe, A. G. Vitenberg, A. N. Marinichev, and L. M. Kuznetsova, *Chromatographia*, **9**, 502 (1976).
41. B. V. Ioffe, A. G. Vitenberg, A. N. Marinichev, and L. M. Kuznetsova. *Zh. Prikl. Khim.*, **49**, 1759 (1976).
42. B. V. Ioffe, A. G. Vitenberg, and A. N. Marinichev. USSR Pat. 480, 934; *Byull. Izobr.*, **1976** No. 39.
43. A. G. Vitenberg, B. V. Ioffe, Z. St. Dimitrova, and T. I. Strukova. *Dokl Akad. Nauk SSSR*, **230**, 849 (1976).
44. A. G. Vitenberg, B. V. Ioffe, Z. St. Dimitrova, and T. P. Strukova. *J. Chromatogr.*, **126**, 205 (1976).
45. A. G. Vitenberg and T. P. Strukova, *Koordinats. Khim.*, **4**, 361 (1978).
46. A. G. Vitenberg, B. V. Ioffe, and Z. St. Dimitrova, *Dokl. Bolgar. Akad. Nauk*, **31**, 1023 (1978).
47. A. G. Vitenberg, B. V. Ioffe, and Z. St. Dimitrova, *J. Chromatogr.*, **171**, 49 (1979).
48. J.-C. Leroi, J.-C. Masson, J.-F. Farbles, and H. Sannler, *Ind. Eng. Chem., Proc. Des. Dev.*, **16**, 139 (1977).
49. P. Duhem and J. Vidal, *Fluid Phase Equilibria*, **2**, 231 (1978).
50. E. Santacesaria, D. Berlendis, and S. Carrá, *Fluid Phase Equilibria*, **3**, 167 (1979).
51. Y. Yoshikawa, K. Arita, A. Inaba, and K. Sasaki, *Bull. Chem. Soc. Jap.*, **44**, 2568 (1971).
52. M. O. Burova, A. A. Zhukhovitskii, M. L. Sazonov, and M. S. Selenkina, *Tr. Vses. Nauchno Issled. Geologorazved. Neft. Inst.*, **1973** (112), 105.
53. P. M. Mitruka, *GC Application in Microbiology and Medicine*, Wiley, New York, 1975.
54. L. Larsson and P.-A. Mårda, *Acta Pathol. Microbiol. Scand., Sect. B*, **1977** (Suppl. 259), 5.
55. J. Wüst, *Schweiz. Med. Wschr.*, **110**, 362 (1980).
56. H. Thadepalli and P. K. Gangopadhyay, *Chest*, **77**, 507 (1980).
57. A. J. Taylor, in *Applied Headspace Gas Chromatography* (B. Kolb, Ed.), Heyden, London, 1980, p. 185.
58. L. Larsson, P.-A. Mårdh, and G. Odham, *Acta Pathol. Microbiol., Scand., Sect. B.*, **86**, 207 (1978).
59. L. Larsson, P.-A. Mårdh, and G. Odham, *J. Clin. Microbiol.*, **7**, 23, (1978).
60. N. J. Hayward, T. H. Jeavons, A. J. C. Nicholson, et al., *J. Clin. Microbiol.*, **6**, 187, (1977).
61. P. J. Coloe, *J. Clin. Pathol.*, **31**, 365, (1978).
62. B. Kolb, G. Beyaert, and P. Pospisil, *Applications of Gas Chromatographic Headspace Analysis*, Bodenseewerk Perkin-Elmer GmbH, No. 26, 1980.
63. J. Norrman, *Acta Pathol. Microbiol. Scand., Sect. B,* **1977** (Suppl. 259), 25.

Index

Absolute calibration, 42, 49
Absorber, selection of, 64. *See also* Concentrators
Absorption, atmospheric moisture, 24. *See also* Concentration methods
Accessories, automated, 85–96
Accuracy:
　aqueous solution analysis, 116
　blood alcohol analysis, 119
　blood alcohol determination, 123
　equilibrium concentration, with head-space analysis, 213
Acetaldehyde:
　equilibrium concentration in volatile liquids, 195
　as secondary fermentation products, 147
Acetamide, as polymer solvent, 132
Acetic acid, 58, 59
　distribution coefficients of aromatic hydrocarbons, 199
　equilibrium concentration in:
　　as absorbing liquid, 188–191, 192, 193, 194, 195
　　contaminants, 205–208
Acetone, 62, 192
　in aqueous solution, 117
　in biological tissues, 128
　as blood alcohol standard, 120
　distribution coefficients, 17
　equilibrium concentration of, 175, 178, 195
　as polymer solvent, 133
　in urine, 128–129

Acetylene:
　group reagents, 220–221
　in transformer oils, 157, 158, 160
Acid-base properties:
　gases, 148
　nonseparated bases, 254
Acids, group reagents, 220–221
Acrylates, 134, 135
Acrylonitrile, 135
Activated charcoal, 114, 115
Activity coefficients, 13
　chemical potentials, 11
　determination of, 14–15, 254–256
Additivity rule, mixed solvent composition, 23–24, 25
Adhesive tapes, vapor-phase analysis of, 145
Adsorption:
　analyzed substances, 63
　in blood alcohol analysis, 119
　concentration of substance, 75
　in equilibrium concentration, in nonvolatile liquids, 170–172
　polar substances, 33–34
　sample, 69, 71
　and sensitivity, 57
　in stripping, 109
　Tenax-GC, 111
　in variable volume containers, 75
Air, organic impurities in, 176, 178–179
Alcohols, 58, 62
　in aqueous solution, 117
　bacterial metabolites, 257, 258

265

Alcohols (*Continued*)
  in blood and urine, 117–126
  from fungi, 261
  group reagent, 220–221, 222, 223
  sodium nitrite and, 218
  Tenax and, 112
Aldehydes:
  group reagents, 220–221, 223
  and odors, 233
  Tenax and, 112
Aliphatic amines, 254
Alkaline hydrolysis, chloral hydrate, 130
Alkylnitrites, sodium nitrite and, 218
All-Union Energy Institute, 155
Amines, 69
  bacterial metabolites, 258
  ionization constant determination, 247–254
  microorganism metabolites, 257
  and odors, 233
Anaerobic bacteria, identification of, 258, 259
Angle coefficient(PSI), 30
Aniline, ionization constant, 250
Apiezon, 179
Apiezon K, 178
Apiezon L, 177, 193
Apparatus, *see* Equipment; Instrumentation
Aqueous solutions:
  additivity rule, 24
  analysis of, 100–117
    benzene and toluene in, 102, 103, 104
    seawater, hydrocarbons in, 115
    standards, 114–115
    stripping, 106–112, 114
    sulfur compounds, 101
  and dynamic systems, 83
  mineralization of, 62
  see also Equilibrium concentration, in volatile liquids
Arabinose, 260
Argon-ionization, 235
Aromatic hydrocarbons:
  in biological tissues, 128
  enrichment of, 60, 61
  equilibrium concentration, 175
  equilibrium concentration in volatile liquids:
    absorbers, 192, 199–204
    with varying distribution coefficients, 205–208

group reagents, 220–221, 223
in water, 101–117
Atmosphere, trace contaminants, 2, 176, 178–179
Atmospheric moisture, absorption of, 24
Automation, 3
  accessories:
    for Carlo Erba apparatus, 85–86
    for Hewlett-Packard apparatus, 88–90
    for Perkin Elmer apparatus, 90–96
  equilibrium concentration, with head-space analysis, 213
  extractions of polystyrene sample, 144
Azeotropes, 54, 240, 241, 242

Bacteriology, HSA applications, 257–261
Barley, 147
Bases, volatile organic, ionization constants of, 247–254
Benzene, 36, 59, 192, 219
  in air, 178, 180
  distribution coefficients, 16
  equilibrium concentration of:
    in nonvolatile liquids, 176, 177
    solvent impurities and, 207–208
    in volatile liquids, 194, 195, 199–204
  equilibrium concentration in, 188–191, 195–196, 197
  in water, 101, 102, 103, 104
Biological materials, 4
  head space analysis applications, 257–261
  organic substances in, 62
  Perkin Elmer analyzers, 96
  vapor phase analysis, 225–229
  volatile organic substances in, 117–130
Blood:
  alcohol content, 117–126
  carboxyhemoglobin in, 129
  Perkin Elmer analyzers, 96
  vapor phase analysis, 225–229
Boiler water, hydrogen in, 5, 155
Boiling, during stripping, 107
Brandy, 230
Bromination, 220
Bubbler, extraction of gases, 159
Bunsen coefficients, 148
Butadiene, 132
Butadiene rubbers, gasses from, 145
Butanol, 110

distribution coefficients, 17
equilibrium concentration in, 188–191
secondary, in blood and urine, 126–127
tertiary, in blood alcohol determination, 120, 121, 122, 123, 124
n-Butyl acetate, distribution coefficients, 17
Butyl rubber septa, 92, 93

Calcium hydride, as group reagent, 218, 221
Calibration:
  absolute, 49
  for aqueous solutions analysis, 106
  for blood alcohol analyzer F-45, 123–124, 125
  detector, 4
  in equilibrium concentration, in nonvolatile liquids, 171
  as head-space analysis application, 234–242
  standards, 40–45
Calibration factors, biological fluid analysis, 125
Carbon dioxide:
  and degradation of sample, 234
  dissolved, 154, 156
  in transformer oils, 157, 158–159, 160
Carbon disulfide, 113, 114
Carbon monoxide:
  dissolved, 154, 156
  in transformer oils, 157, 159
Carbonyl compounds:
  in barley and malt, 147
  equilibrium concentration in volatile liquids, 192, 195, 199–204
  microorganism metabolites, 257
  see also Hydrocarbons; specific compounds
Carboxyhemoglobin, in blood and tissue, 129
Carlo Erba apparatus, 85, 147
Carotenoids, extraction of, 147
Carrier-gas flow, 75
Catharometer, 129
Caviar, odors, 233
Celite 545, 175, 177, 178
Cheese, 147, 224
Chemical potential, 10–11
Chloralhydrate, in blood and urine, 129–130
Chlorinated aliphatic hydrocarbons, in biological tissues, 128
1-Chloroalkanes, 114–115

Chlorobenzene:
  in air, 178
  calibration with, 237
Chloroform:
  calibration with, 237
  from chloral hydrate, 130
Chromadistillation, 108
Chromato-distributive method of analysis, 216
Chromatographic column:
  automated injection to, 90, 91
  sample collection from, 82
Chromatographic separation, oxygen-containing hydrocarbons, 204
Chromosorb, 193, 218
Coefficient of proportionality, 11
Coefficients, extraction, dissolved gases, 160
Coffee, odor of, 232, 233
Complex ions, stability constants, 243–247
Complex mixtures, chromatographic profile of, 224–234
Composition, equilibrium phases, 10
Concentration factor, in gas distribution law, 148
Concentration methods:
  aqueous solutions, 108–117
  equipment, 83–84
  and stability of vapor concentration, 235
  see also Cryogenic concentration; Equilibrium concentration
Concentrators:
  column material, 175
  elution from, 171–172, 174–175
Condensation:
  in blood alcohol analysis, 119
  impurity concentration, 83
  saturated vapor, 68
Constant gas space volume, 72
Constant volume systems:
  containers, 68
  sampling from, 72–74
Continuous analysis, of dissolved gases, 154
Continuous gas extraction, 49–57
  distribution coefficient determination, 34–37
  sample introduction, 75
Coordination substances, 243–247
Cottonseed oil, 147
p-Cresol, bacterial production of, 260–261

Cryogenic concentration:
  absorber selection, 64
  analyzed substances, 63
  in equilibrium concentration, 179
  equipment, 84
  sample introduction for, 75
  and sensitivity, 57
  after stripping, 115
  in water solution, 108–117
Cyanoethylated pentaerythritol, 204
Cyclohexane, in adhesive tapes, 145

Dairy products:
  odors, 233
  volatile substances in, 218
Degassing, of solutions, 150–153
Dehydration, biological sample, 129
Dependence, functional, 10
Desorb cycle, HP-7675A, 89
Desorption:
  solvents for, 113–114
  in stripping, 109
  thermal, 145, 179
Diabetes, 225, 226, 228, 229
o-Dichlorobenzene, as polymer solvent, 133
Dichloroethane, in corpses, 128
Diethylamine:
  equilibrium concentration, in volatile liquid, 194, 195
  ionization constant determination, 251, 252
  see also Amines
N-Diethylaniline, ionization constant, 250
Diethylene glycol adipate, 179
Diisopropylamine, ionization constant, 250
Diisopropyl ether, calibration with, 237
Dilution, degree of:
  and additivity rule, 23
  and error, 235–236
  and sensitivity, 58–64
  unstable compounds, 234
Dimethylacetamide, as polymer solvents, 133
Dimethyl disulfide, bacterial metabolites, 258
Dimethylformamide, 58, 132, 133
Dimethyl sulfide:
  distribution coefficients, 17
  equilibrium concentration, 195
Dimethylsulfoxide, 133
Dioxane:
  as blood alcohol standard, 120
  distribution coefficients, 17, 18–21
Direct vapor-phase analysis, of polymers, 139–141
Discontinuous gas extraction, sampling system, 143
Dissociation, gases, 148
Distillation:
  biological samples, 127–128
  stripping by, 107–108
Distilled water, chromatograms of, 194
Distribution, maximum, 14
Distribution coefficients, 11, 14
  in dilute solutions, 235–236
  equilibrium concentration in nonvolatile liquids, 171
  equilibrium concentration in volatile liquids, 187, 192
    aromatic hydrocarbons, 199, 200, 201, 202, 203
    with head space analysis, 209–213
    varying values, 204–208
  in equilibrium constant determination, 243
  gases, 150
  group identification by, 216–217
  lowering, 57–58
  measurement of, 27–37
    continuous gas extraction, 34–36
    errors, sources of, 29, 30, 33–34, 35
    fixed-volume systems, 31–33
    graphical methods, 30–31
    indirect method, 27–28
    lower limits, 29
    low values, 28
    medium values, 31
    multiple extraction, 30–31
    substitution method, 28–31
    upper limit, increasing, 30
  mixed solvent composition, 23–24, 25
  monomers, 134–135
  of organic substances, 16–17
  and sensitivity, 61–64
  unknown, 4
  in variable volume containers, 75
  volatile absorbing liquids, 183–184
  in water analysis, 102
Distribution laws, 11
Dynamic systems:
  methods, 4, 25, 49–57
  sample introduction, 75, 81

INDEX 269

E-301 silicone, 177, 180
Ecological chemistry, dissolved gas determination, 156
Electroinsulating oils, gases in oil of, 156, 157, 158–160
Electrolytic dissociation, gases, 148
Electron capture, 92, 147
Elution, from concentrator, 171–172, 174–175
Emulsions:
  and error, 60–61
  polymer, 134–136
Enrichment:
  analyzed substances, 63
  see also Concentration; Cryogenic concentration; Equilibrium concentration
Entrainment separator, 75
Environmental media, Perkin Elmer analyzers, 96
Equilibration, degrees of, 2–3
Equilibrium:
  phases, concentrations in, 10
  sample introduction and, 68
  in variable volume containers, 74–75
Equilibrium concentration, 5–6
  in nonvolatile liquids, 168–169, 170–180
    atmospheric impurities, 178–180
    techniques, 170–178
  in volatile liquids:
    accuracy, 194–195
    aromatic hydrocarbons, 199, 200, 201, 202–204
    impurity concentration curves, 182–183
    liquids, properties of, 187–188, 189
    maximum values, 185–186, 189, 192
    methods, 169
    purging, 180–181, 182
    sensitivity, 186–187
    sulfur-containing substances, 192, 193, 195–198, 199
    theory, correlation with, 188–189, 190–191, 192
    volatility and, 183–184, 189
    see also Aqueous solutions
Equilibrium constants, organic substances, 242–254
Equilibrium vapor analysis, of blood, 124–128
Equipment, 75–84
  sampling, 79–83
  systems, 75–79

see also Instrumentation
Error sources:
  in distribution coefficient determination, 29–30
  adsorption and, 33–34
  in continuous extraction, 35, 36
  emulsions as, 60–61
  in equilibrium concentration in volatile liquids, 185
  in ionization constant determination, 247–254
Escaplen, 138
Escherichia coli, 260
Esters:
  from fungi, 261
  group reagents, 220–221, 223
Ethane, in transformer oils, 157, 158, 160
Ethanol, 62, 222
  blood content, 117–126
  distribution coefficients, 17, 18–21
  in urine, 128–129
Ethers:
  group reagent, 223
  Tenax and, 112
Ethyl acetate, 62
  distribution coefficients, 17
  as secondary fermentation products, 147
Ethylbenzene, distribution coefficients, 16
Ethylene:
  as pyrolysis product, 156
  in transformer oils, 157, 158, 160
  vapor phase identification, 218
Ethylene glycol, 36
Ethylene oxide, 138–139, 143
2-Ethylhexyl acrylate, 132
Ethyl mercaptans, distribution coefficients, 17. See also Mercaptans; Sulfur-containing compounds
Ethyl nitrite, 222
Evaporation:
  and impurity concentration, 83, 182, 183–184, 189
  volatile compounds, 141. See also Equilibrium concentration, in volatile liquids
Extraction methods, 5
  aqueous solutions, 108–117
  of carotenoids, 147
  continuous, 3
  distribution coefficient determination, 34–37

Extraction methods (*Continued*)
  gases in solutions, 150–154
  multiple-stage, 45–47
  polymers, 139–145
  single-stage, 37–45
  in stripping, 109
  see also Gas extraction
Extraction coefficients, dissolved gases, 160

F-40, 135
F-45, 90, 91, 92, 123
Factors, calibration, 125
Fatty acids:
  bacterial metabolites, 258
  microorganism metabolites, 257
Fermentation, secondary products, 147
f function, 10
Films, 138
Filters, hydrophobic, 83
Fingerprints, vapor phase, 226, 227
Fish, odors, 233
Fixed-volume systems, distribution coefficient determination, 31
Flame-ionization detector, 177, 178, 193, 194
  aqueous solution analysis, 100, 101, 104
  equipment, 92
Flame-photometric detector:
  calibration of, 235
  equipment, 92
  sulfur-containing compounds, 195, 199
Flow rates, and equilibrium concentration, 188
Flow-through microreactor, 218
Fluoroplastic glass, 83
Food products:
  analysis of, 146–148
  cryogenic trapping, 106
  odors of, 230–231, 232–234
  Perkin Elmer analyzers, 96
  vapor phase group identification, 218
Foreign substances, and distribution coefficient, 24
Forensic chemistry, 117
Forosilicate glass, 83
Freons, in sea water, 156
Functional dependence, 10
Fungi, metabolites, 261
Fusel oil alcohols, 147

Gaseous media, trace contaminants, 2

Gases:
  from polymers, 145
  in solution, 148–160
    extraction methods, 150–155
    gas laws, 148, 149, 150
    in steam boilers, 155–156
    in transformer oils, 156, 157, 158–160
    see also Equilibrium concentration
Gas extraction:
  activity coefficient determination, 254–256
  continuous, 3
  distribution coefficient determination, 34–37
  methods of, 5
  multiple-stage, 45–47
  polymer, 139–145
  single-stage, 37–45
Gas flow rates, and equilibrium concentration, 188
Gasoline:
  analysis of, 115
  vapors, in air, 180
Gas solubility coefficients, 148, 149
Gas syringe, sampling with, 69
Glass beads, 230
Goke method, 77–78
Gottauf technique, 104
Group reagents, 217, 218, 220–221

Halcomide, 18, 122
Halothane, 178
Head-space analysis, with equilibrium concentration, 209–213
Helium:
  dissolved gas extraction, 153–156
  purging with, 104, 227
  see also Inert gases
Henry's coefficients, 23–24
Henry's constant, 148, 235
Henry's law, 12–13
Heptane:
  calibration with, 237
  sulfuric acid and, 219
n-Heptylbenzene, 109
Heterocyclics, ionization constant determination, 247–254
Hewlett-Packard equipment, 88–90, 112–113
Hexane:
  in adhesive tapes, 145
  sulfuric acid and, 219

HP-7675A, 88–90
HS-6, 86–88
HS-250, 85–86
Hydrazines, ionization constant determination, 247–254
Hydrocarbons, 69
  concentrator column material, 175
  distribution coefficients, 16–17
  in oils, 155, 157, 158–160
  sulfuric acid and, 219, 220
  see also Aromatic hydrocarbons; *specific substances*
Hydrogen:
  in boiler water, 5
  dissolved, 153–154, 155
  as group reagent, 218, 220
  in transformer oils, 157, 158, 159, 160
Hydrolysis:
  chloral hydrate, 130
  and degradation of sample, 234

Illuminating gas, trace components in, 179
Impurities:
  accumulation of, in volatile liquids, 181
  concentration of, 83–84. *See also* Concentration methods; Cryogenic concentration; Equilibrium concentration
  and distribution coefficient, 24
Incomplete gas extraction, dissolved gases, 160
Inert gases, 77
  activity coefficient determination, 254–256
  dissolved gas extraction, 153
Infection, diagnosis of, 258–260
Influenza, 229
Injection:
  equipment for, 79–83
  pneumatic, 71–73
  see also Sample, introduction of
Instrumentation:
  automated accessories, 85–96
  biological sample analysis, 128–129
  gas extraction, 151–155
  group identification, vapor phase analysis, 217
  hydrogen in solution, 155
  introduction of vapor phase methods of, 68–75

laboratory equipment, 75–84
  see also Equipment
Insulation, pyrolysis of, 156
Insulation oils, dissolved gases in, 155, 156, 157, 158–160
Internal standards, 43–45. *See also* Standards
Ionization constants:
  organic bases, volatile, 247–254
  organic substances, 242–254
Ionization detector, 92
Isobutyl alcohols, in blood and urine, 126–127
Isopropanol:
  as blood alcohol standard, 120
  sodium nitrite and, 222
  in urine, 128–129

K, *see* Distribution coefficients
Katharometer, 92
Kefir, volatile substances in, 218
Ketones:
  cyclic, 231
  group reagent, 223

Latexes, volatile components of, 134
Law of distribution, 11
Leakage, sample, 69, 71
Leucine, 261
Linde molecular sieve, 154
Liquid-gas systems, distribution coefficient measurement, 18–19, 27–37
Liquids, addition to constant volume containers, 73. *See also* Equilibrium concentration

Malt, 147
Mass balance, equation of, 181
Medical diagnostics, head-space analysis in, 257–261
Medium pressure polyethylene(MPE), 138
Mercaptans:
  distribution coefficients, 17
  equilibrium concentration, 195
  see also Sulfur-containing compounds
Mercury pumps, gas extraction, 151–155
Metabolites, volatile, 257
Meteorology, head-space analysis, 237
Methacrylate, 146
Methane:
  dissolved, 154, 156

INDEX

Methane (*Continued*)
  as pyrolysis product, 156
  in transformer oils, 157, 158, 159–160
  vacuum extraction of, 159
Methanol:
  equilibrium concentration, 175, 178
  sodium nitrite and, 222
  in urine, 128–129
Methionine, metabolism of, 260
Methods, *see* Theory
Methyl butyl ketone, in aqueous solution, 117
Methylcyclopentane, in adhesive tapes, 145
Methyl-dipropylamine, 251, 252
Methylene chloride, 114
Methyl ethyl ketone, 192
  as blood alcohol standard, 120
  distribution coefficients, 17, 18–21
  equilibrium concentration, in volatile liquid, 194, 195
Methyl mercaptan, distribution coefficients, 17. *See also* Mercaptans; Sulfur-containing compounds
Methyl methacrylate, 146
2-Methylnaphthalene, 109
Methyl nitrite, 222
Methyl propyl ketone:
  distribution coefficient, salts and, 62
  equilibrium concentration in volatile liquids, 195
Microbiology, head-space analysis in, 257–261
Micro-coulometric detector, 101
Microorganisms, in blood alcohol analysis, 119
Microreactor, flow-through, 218
Minerals, in stratal water, 102
Molecular sieve(s):
  gas trapping, 154
  vapor phase analysis, 218, 221
Molecular weights, determination of, 256–257
Mole fractions, chemical potentials, 11
Multicraft F-40, 135
Multicraft F-45, 90, 91, 92, 123
Multiple gas extraction, 5, 45–47
Musk, 231

Naphthalenes, equilibrium concentration in volatile liquids, 204
Natural waters:
  atmospheric gases in, 155
  stratal, 101–102
Nitrogen, dissolved, 153–154
Nitrogen heterocyclics, ionization constant determination, 247–254
Nitromethane, distribution coefficients, 18–21
Nonane, 219
Nonseparated peaks, analysis of, 253–254

n-Octane, distribution coefficients, 18–21
Odors, analysis of, 229–234
Oil deposits, water analysis, 101–104
Oils:
  dissolved gases in, 156
  hydrocarbons in, 155
  outgassing, 151
  pyrolysis of, 156
  in transformers, 156, 157, 158–160
O.K.B. Automatiki, 152
Olefins, group reagents, 220, 223
Organic liquids, and dynamic systems, 83
Organic substances:
  bases, ionization constants of, 247–254
  distribution coefficients of, 16–17
  ionization constant determination, 242–254
  solvents, and sensitivity, 61–62
  see also Aliphatic hydrocarbons; Hydrocarbons; *specific compounds*
Organs, sample preparation, 127–128
Ostwald coefficient, 148
Outgassing, dissolved gases, 150–154
Oxidation, sample degradation, 234
Oxygen, dissolved, 153–154
Oxygen-containing compounds:
  distribution coefficients, 17
  equilibrium concentration in volatile liquids, 204
  group reagents, 220–221
Ozone, as group reagent, 218, 221

Packaging materials, impurities, 146, 147
Paraffins:
  equilibrium concentration in volatile liquids, 204
  group reagent, 220, 223
Paraldehyde, in blood and urine, 129–130
Partition coefficients, unknown, 5
Pauschmann method, 76–77
Peak areas:
  calculating, 142–143

ionization constant determination, 247–254
  organic bases, 253
Pentane, trichloroethylene extraction, 148
1-Pentanol, 110
Penterythritol, cyanothylated, 204
Perkin Elmer apparatus, 72, 86–88, 90–96
  aqueous solutions, 100–104
  blood alcohol analysis, 118, 125
  equilibrium concentration, with head-space analysis, 213
  polymer analysis, solution, 134, 135, 136
Phase composition, 10
Phase volume ratio, gases, 150
Phenol, bacterial production of, 260–261
Phosphorus, aqueous solutions, 100
Piorr apparatus, 77–79
Piperidine, 251, 252
Pneumatic injection, 71–73
Pneumatic sampling, 93–95
Polar substances, distribution coefficients, 33
Polyacrylate dispersions, 134
Polyethylene, 138, 218
Polyethylene glycol-300, 36
Polyethylene glycol-400, 175, 180
Polyethylene glycol-450, 178
Polyethylene glycol-600, 193
Polyethylene glycol-1500, 122
Polymers, volatile substances in, 131–146
  emulsions, 134–136
  films, 138
  solids, 136–146
  solutions, 132–134
Polystyrene, 143, 145, 146
Polyvinyl acetate, 134
Polyvinyl chloride, 145
  solvents, 133
  vinyl chloride analysis, 137, 147
Porapak, 122, 178, 179, 233
Porasil, 227, 228
Potassium hydroxide, 192, 194
Pressure, 10
  and equilibrium, 68
  in sample introduction, 68, 69, 70, 71
  sampling and, 73–75
    constant, 79–80
    equilization of, 78–79
Propane, in transformer oils, 157
Propanol:
  as blood alcohol standard, 120
  in blood and urine, 126–127
  distribution coefficients, 17
  sodium nitrite and, 222
Proportionality coefficient, 11
Propylene, in transformer oils, 157
Propyl nitrite, 222
Proteus mirabilis, 260
Purge cycle, HP-7675A, 89
Purging:
  helium 104, 227
  impurities in gases, 204–205
  mercaptans, 196–197
Purging volume, in equilibrium concentration, in nonvolatile liquids, 171
Pyridine, 250, 251, 252, 254

Qualitative analysis, 83
  complex mixtures, 224–234
  equilibration in, 2n.
  individual and group identification, 216–224
  odors, 229–234
Quantitative analysis:
  food products, 146–148
  gases in solution, 148–160
  volatile organic substances:
    in biological systems, 117–130
    in polymers, 131–146
  water and aqueous solutions, 100–117

Raoult's law, 24, 256, 257
Reagents, group, 217, 218, 220–221
Reproducibility:
  blood alcohol analysis, 119
  determination of, 236–240
  sample introduction and, 69–70
Retention times, of toxicological substances, 122
Retention volume, in equilibrium concentration, 171–172
Reverse head-space analysis:
  characteristics and type of, 168–170
  in nonvolatile liquids, 170–180
  in volatile liquids, 180–204
    with equilibrium concentration, 209–213
    with variable distribution coefficients, 204–208
Rubber:
  absorption by, 69
  butadiene, 145

Rubber (*Continued*)
  contaminants, 139, 143
Rubber septa, types of, 92–93

Salting-out effect, 24
  blood alcohol analysis, 119
  reagents, for volatile substances, 104–105
Salts, and distribution coefficient, 61, 62
Sample:
  introduction of, 68–75
    pneumatic injection, 71–73
    sampling value, 69–71
    syringe, 69
  preparation of:
    biological tissue, 127–128
    Perkins Elmer apparatus, 93
  volume, blood, 118
Sampler, HP 7675A, 112–113
Sampling:
  from constant volume containers, 72–74
  discontinuous gas extraction, 143
  equipment for, 79–83
    constant pressure, 79–80
    pressure equalization, 78–79
    variable gas phase volume, 77–79
  Perkins Elmer apparatus, 93–95
  repeated, 73
  syringe for, 69
  from variable volume containers, 74–75
Sampling valve, 80–81
Saturated vapor, condensation of, 68
Saturation curves, sulfur compounds, 195
Seawater:
  Freons in, 156
  hydrocarbons in, 115
Sensitivity:
  biological product analysis, extraction and, 129
  equilibrium concentration:
    with head space analysis, 209–213
    in nonvolatile liquids, 173, 174, 177–178
    in volatile liquids, 186–187
  ethylene oxide determination, 138
  impurity concentration and, 83–84
  increasing, 57–64
  polymer analysis, 132, 134
  single-stage extraction, 37–40
Septa, types of, 92–93
Sewage waters, 4, 102

Silicone elastomer E-301, 175
Silicone rubber septa, 93
Single-stage extraction, 37–45
Sodium, as group reagent, 218, 220
Sodium fluoride, in blood alcohol analysis, 119
Sodium nitrate, in blood alcohol analysis, 119
Sodium nitrite, as group reagent, 218, 220
Solenoid valve, 144
Solubility:
  lowering, 58
  and sensitivity, 62
Solubility coefficients, gas, 148, 149
Solution, volatile substances, 56–57
Solutions:
  calibrations, 40–42
  gases in, 148–160
    extraction methods, 150–155
    gas laws, 148, 149, 150
    in steam boilers, 155–156
    in transformer oils, 156, 157, 158–190
  polymer, 132–134
  see also Equilibrium concentration
Solvents:
  in adhesive tapes, 145
  for desorption, 113
  distribution coefficients in liquid-gas systems at 25oC, 16–21
  for equilibrium concentration in volatile liquids, 187–188
  mixed, additivity rule, 23–24, 25
  for polymers, 132, 133
  removal of, 56–57
  volatile, continuous gas extraction of, 52–57
Sorbent, in equilibrium concentration, 170–171
Sorption concentrator, 108
Stability, of vapor concentration, 235
Stability constants, complex ions, 243–247
Standard gaseous mixtures, preparation of, 236–242
Standardization, 10
  in blood alcohol analysis, 120, 121, 123
  calibration, 40–45
  internal, 43–45
  polymer analysis, 138, 139
Static methods, 25
  addition of sample, 47–49
  multiple gas extraction, 45–47

INDEX

sample introduction, 68
single-stage extraction, 37–45
systems, 73
Steam boilers, dissolved gas analysis, 155
Steam distillation, biological samples, 127–128
Stereoisomers, odors of, 231
Stratal waters, 101–102
Stripping:
  dissolved gases, 153–154, 155–156
  seawater, 115
  water samples, 106–112
Structure, and odor, 231
Styrene, 135
  impurities, 143, 146
  solvents, 133
  water and, 132
Sulfur-containing compounds:
  in aqueous solutions, 101
  equilibrium concentration in volatile liquids, 195–198, 199
  microorganism metabolites, 257
  and odors, 233
  in urine, 225
Sulfuric acid, as group reagent, 219, 220
Sulfurous compounds, distribution coefficients, 17
Surface tension, and dynamic systems, 83
Switching system, 76
Syringe, sampling with, 69

Techniques, *see* Theory
Teflon-coated rubber septa, 93
Temperature, 10
  in blood alcohol determination, 119, 121
  and distribution coefficients, of organic substances, 16–17
  sample introduction and, 69
  during stripping, 107
Temperature coefficient, relative, 15
Tenax, 111, 179, 225, 227, 228, 230
Testing:
  gas chromatograph detectors, 237
  head-space analysis application, 234–242
Theory:
  addition of unknown, 47–48
  continuous extraction, 49–57
  distribution coefficient measurement, 27–37
  dynamic methods, 49–57
  principles, 10–26

sensitivity, increasing, 57–64
static methods, 37–49
Thermal desorption, 112, 145, 179
Thermal methods, of increasing sensitivity, 58
Thermodynamic conditions, of phases, 10
Thermodynamic equilibrium, 2–3
Thin films, polymer analysis, 136–137
Tissue, biological, 4
  carboxyhemoglobin in, 129
  sample preparation, 127–128
Toluene, 59, 143, 192, 219
  in air, 180
  calibration with, 237
  distribution coefficients, 16, 18–21
  equilibrium concentration, 176, 177, 178
  solvent impurities and, 207–208
  in volatile liquids, 195, 199–204
  in food containers, 147
  in water, 102, 103, 104
Toxicology, 117
  retention times, 122
  see also Biological substances; Tissue, biological
Trace contaminants, 2
Transevaporator, 226, 227
Transformers, gases in oil of, 156, 157, 158–160
Trap-cleaning cycle, HP-7675A, 89
Trichloroacetic acid, in blood and urine, 129–130
Trichloroethylene:
  in blood and urine, 129–130
  carotenoid extraction, 147–148
Triethylamines, 250, 251, 252
Tris(cyanopropyl)amine, 204
Tyrosine metabolism, 260

UDZh-64, 152
Unsaturated compounds, group reagents, 220–221
Unstable compounds:
  analysis of, 181
  dilute solutions of, 234
  equilibrium concentration and, 169–170
Urine:
  alcohol in, 118
  bacterial identification, 260
  Perkin Elmer analyzers, 96
  vapor phase analysis, 225–229

Urine (*Continued*)
  volatile product analysis, 129–130
USSR, forensic practices in, 128, 129

Vacuum extraction, dissolved gases, 150–153, 159
Valves, sampling, 69–71, 80–81
Vapor concentration, stability of, 235
Vaporization chambers, polymer contamination, 132
Vapor-phase analysis, 217
  of adhesive tapes, 145
  odors, 229–234
  reagents for, 220–221
Vapor-phase concentration, 112
Vapor pressure, in blood alcohol determination, 121
Variable-volume apparatus:
  containers, equilibrium in, 74–75
  system, 77–79
Vials:
  alcohol determination, 129
  for biological materials, 258
  sample, 92
Vinyl chloride, 137
  monomeric, 131, 132
  as polymer solvents, 133
  in vodka, 147
Volatile ligands, complexing of, 243–247
Volatile product analysis, of blood and urine, 129–130
Volatile substances:
  in biological systems 117–130
  continuous extraction from, 52–57
  equilibrium concentration in, *see* Equilibrium concentration, in volatile liquids
  in polymers 131–146
  quantitative determination, 56
  removal of, 56–57
  see also Qualitative analysis; Quantitative analysis; *specific substances*
Volatility, and equilibrium concentration, 183–184, 189
Volatilization, from polymers, 145
Volume(s):
  constant gas space, 72
  in equilibrium concentration:
    in nonvolatile liquids, 171, 172
    in volatile liquids, 188
  gas, in continuous extraction, 51–52
  of sample, blood, 118
  and sensitivity, 59–64
  and stability of vapor concentration, 235

Water:
  analysis of, 100–117. *See also* Aqueous solution
  atmospheric gases in, 155
  atmospheric moisture, 24
  chromatograms of, 194
  equilibrium concentration in, 188–191, 192, 193, 194, 195
  Freons in, 156
  gas solubility in, 149
  as impurity, in equilibrium concentration, 207
  and polymer analysis sensitivity, 132, 134
  waste, 101

m-Xylene, 59, 192
  equilibrium concentration:
    solvent impurities and, 207–208
    in volatile liquids, 195, 199–204
  distribution coefficients, 16
p-Xylene, equilibrium concentration, 176, 177